Making Sense
and
Making Decisions

An engineer's guide to project decision making

Second Edition

AF812420

Howard Duhon, P.E.

Publisher: GATE

Copyright © 2012 by Howard Duhon, P.E.

All rights reserved. No part of this publication may be reproduced, stored in a retrieval system, or transmitted in any form or by any means electronic, mechanical, photocopying, recording or otherwise without prior written permission of the publisher.

Published by:
Gibson Applied Engineering and Technology, Inc (GATE)
Houston, TX
www.GATEINC.com

First Edition Published by:
Decision Engineering
New Orleans, LA

First Edition – December 2006 (paperback)
ISBN No. 0-9771012-0-7

Second Edition – July 2012 (paperback)
ISBN 13. 978-0615667720

Cover design by Stephanie Walsh and Howard Duhon

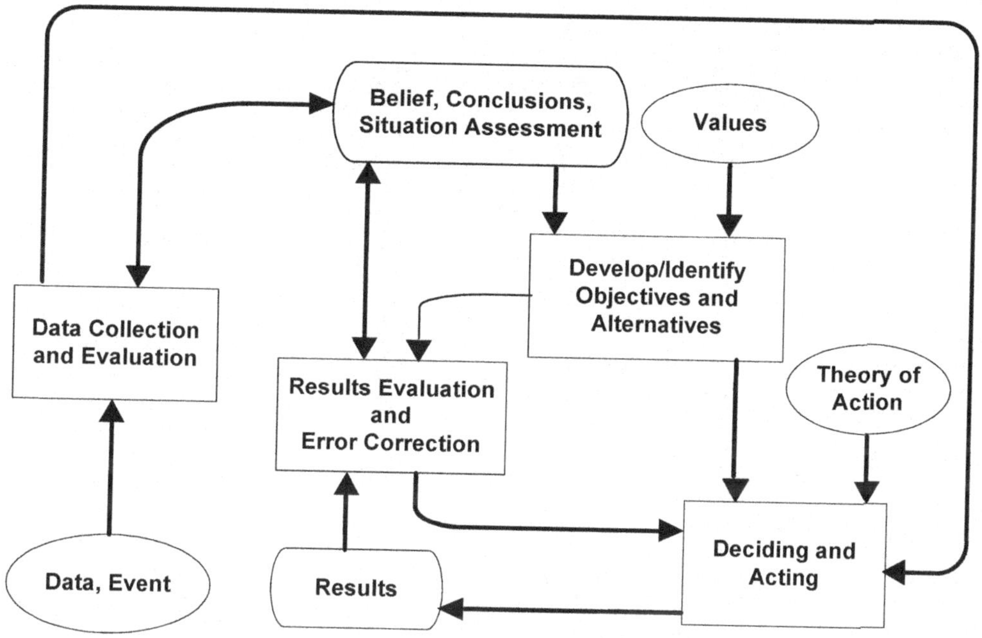
Belief, Conclusions, Situation Assessment
Values
Develop/Identify Objectives and Alternatives
Data Collection and Evaluation
Results Evaluation and Error Correction
Theory of Action
Data, Event
Results
Deciding and Acting

Engineering is:

the use of heuristics

to find the best solution

to an ill-defined problem

subject to resource limits.

Paraphrased version of the Engineering Method from *Discussion of the Method* by Billy Vaughn Koen (33).

Heuristic:

1) relating to general strategies or methods for solving problems, especially of a method that is not certain to arrive at an optimal solution
(From Wiktionary)

2) rule-of-thumb
(common use)

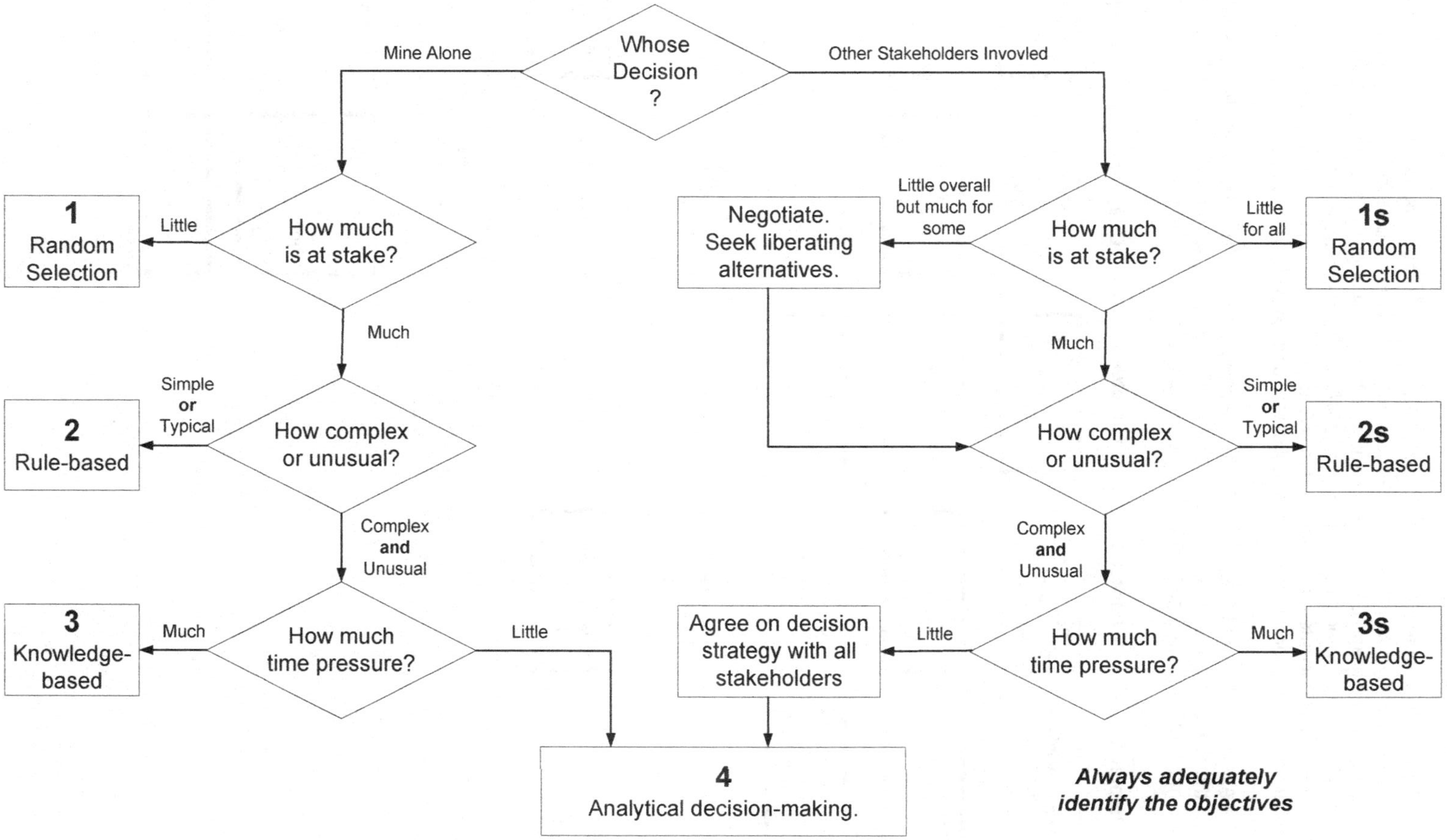
Whose Decision ?
Mine Alone
Other Stakeholders Invovled
How much is at stake?
Little
1 Random Selection
Much
How complex or unusual?
Simple or Typical
2 Rule-based
Complex and Unusual
How much time pressure?
Much
3 Knowledge-based
Little
How much is at stake?
Little for all
1s Random Selection
Little overall but much for some
Negotiate. Seek liberating alternatives.
Much
How complex or unusual?
Simple or Typical
2s Rule-based
Complex and Unusual
How much time pressure?
Much
3s Knowledge-based
Little
Agree on decision strategy with all stakeholders
4 Analytical decision-making.
Always adequately identify the objectives

HOW TO READ THE BOOK

This book is written as the textbook for a seminar course. It is designed to be fairly comprehensive. It is not light reading material. Some topics are more important than others. The graphic below (this page and the facing page) summarizes course topics with the more important section in bold. While the chapters are largely standalone there are concepts that build gradually over the course. In particular, the concept of theories of action in Chapter 5 is critical for understanding Chapter 6.

All chapters except 1 and 9 end with a summary section and then a section answering the four questions. It may be useful to read those two sections first, then decide how much effort to put into the details of each chapter.

Chapter 1	Chapter 2	Chapter 3	Chapter 4	Chapter 5
Preliminary Info 1.1-1.5	Reasons for belief 2.2, 2.12	Value-focused thinking 3.1-3.3	Probability, Statistics 4.2, 4.3, 4.5-4.10	Actionable and Effective plans 5.1 - 5.7
Objectives of the course 1.6	Biases and Heuristics 2.3 - 2.10	Structuring objectives 3.4	Bayesian Reasoning 4.4	Planning Fallacy 5.8
Organization of the course 1.7, 1.8	Coping with emotions 2.11	Identifying Objectives 3.5-3.8	Distributions 4.8 - 4.11	Human Error 5.6, 5.7 5.9 - 5.12
	Selecting and evaluating data 2.13 - 2.14	Use Cases 3.6	Monte Carlo Simulation 4.12	Systems Theory 5.13
	Sensemaking 2.15 - 2.20	Attributes (measures of success) 3.9 - 3.13	Optimization 4.13 - 4.15	**Theory of Action Basics 5.14 - 5.16**
		Identifying Alternatives 3.14	Decision trees 4.16	Additional Theory of Action 5.17 – 5.28
		Influence Diagrams 3.15	Preference Theory Utility Theory Propsect Theory Regret Theory 4.17 - 4.23	Influence 5.29, 5.30 Emotions 5.31 - 5.34
			Decision strategies 4.24 - 4.27	Identifying Bad Advice 5.36

Chapter 6	Chapter 7	Chapter 8	Chapter 9
Single-loop vs. Double-loop learning 6.1 - 6.3	Stress 7.2, 7.3, 7.8, 7.9	Team basics 8.1 - 8.3	Reasons for **Late Changes** summary and suggestions 9.1
Learning case studies 6.4 - 6.10 **6.11 - 6.14** 6.15 6.17	Recognition Primed Decision model (RPD) 7.4 - 7.7	Team competence 8.4, 8.5	Reasons for **Interface Problems** summary and suggestions 9.2
Organizational Learning 6.16	Sources of power 7.10 - 7.13	Negotiang Art of Rhetoric 8.6	Reasons for **Repeating Mistakes** summary and suggestions 9.3
	Cognitive Task Analysis (CTA) 7.14 - 7.16	Interfaces between teams 8.6 - 8.12	Reasons for **Schedule Failure** summary and suggestions 9.4
			Suggestions Summary 9.5
			Decision Strategy 9.6 – 9.12

ACKNOWLEDGMENTS

The book began life as notes for a seminar course. I gave the course several times before deciding that it should be a book. The first three courses were delivered to friends. Those early courses helped me tremendously in refining the work. For braving those early trials and offering helpful advice I'd like to thank: Bosun Oni, Paul Paterson, Susan Lorimer, Andrew Kaczmarski, Susan Lindsey, Vicki Niesen, Jonny Gibson, Lee Jordan, Mark McKay, Jerry Jackson, Lindsay Anderson, Mark Gregory, David Fleming, Ravi Nath, Ken Schroeder, Todd Barlow, Gary Killian, Justin Landry, Mike Walsh, Janet Elias, Jim Hooton, Kelda Charman.

For their interest in this undertaking and for many enlightening conversation and helpful suggestions I want in particular to thank Paul Paterson, Jeff McPhate, Janet Elias and John Walsh.

Thanks to Stephanie Walsh for her help in designing the book cover.

A special thanks to Paul Paterson for his meticulous proof reading of the first edition.

And finally, a special thanks to my loving wife Thais for her patience with this project and her many helpful insights along the way.

Questionnaire

1) In general we make better business decisions when we keep emotions out of it. True or False ____________

2) Do you make more conscious or unconscious decisions? ____________

3) Two people in your firm take a random drug test. One is a middle aged accountant with a boring life. The other is a 20 year old motorcycle rider with tattoos, an earring and a propensity for all night drinking binges. Both tests come back positive for marijuana. The test is 99% accurate.

   What is the probability that the accountant is a user? ______

   What is the probability that the biker is a user? ______

   Do you believe that it is equally likely that they are both users? ______

4) The more important a decision is, the more important it is to collect a great deal of data about each of the alternatives and to evaluate that data in meaningful ways, i.e. via statistical regression or via spreadsheet comparisons? True or False ____________

5) If your answer to question number 4 is "true", did you use such a technique to pick your spouse? Do you still have the spreadsheet? And did your spouse compare you to other potential mates via a structured decision analysis tool?

6) a) Suppose that you are responsible for carrying out a dangerous task. There is a 1 in 1000 chance that you will be killed doing it. Danger is the only downside; it is not demeaning, uncomfortable, painful, illegal, immoral, etc. Suppose that someone has offered to take this task off your hands for a fee. How much would you be willing to pay to have this burden lifted? ________

   b) Suppose instead that someone else is responsible for carrying out the dangerous task. That person offers to pay you to do it for them. How much would they have to pay for you to take the task on? ________

   c) Are your answers for questions 6a and 6b the same? If not, why not?

7) A rational strategy for data gathering is to collect data until the cost of collecting more data is greater than the worth of the data? True or False __________

8) If you answered "true" to question 7 describe how you would go about determining the value of data that you have not yet collected?

9) A small, cohesive team is debating a controversial issue. Would you expect the team's ultimate decision:

   a) To represent the average of the individual member's positions on the issue
   b) To approximate the most radical position held by any one team member
   c) To be more radical than the position of the most radical team member

10) The answer to question 9 is frequently ‘C’. Why?

11) a) Would you play this game? __________
On the flip of a fair coin: Heads you win $10 and Tails you lose $10.

b) Suppose that I gave you $10 first. Then would you play the game? _______

c) Suppose that you owed me $10. Then would you play the game? _______

12) How does an outfielder (baseball) judge a fly ball? If you were to program a robot to catch fly balls what data would you collect and how would you evaluate the data? Do you think that the human approach to judging a fly ball bears any resemblance to the program you would write for the robot?

13) It has been claimed that playing classical music to unborn fetuses will make them smarter and more creative. Do you believe this? Why? Is this claim something you are likely to believe until proven false or disbelieve until proven true?

14) Investigators have concluded that a piece of steel left on the runway punctured a tire of the Concorde. The metal and/or pieces of the tire were propelled into and ruptured a fuel tank causing the crash. The designers of the Concorde knew that foreign objects on the runway posed a threat; hence there was a procedure to sweep the runway prior to each takeoff. The runway was not swept prior to the ill-fated takeoff. Hence, the cause of the crash was:

- Failure to sweep the runway
- Resulting in projection of object(s) into the fuel tank

The response to the crash by British Air and Air France was to ground the planes for months while the fuel tank was modified to make it more resistant to flying debris.

Why did they redesign the wing rather than simply reinforced the need to sweep the runway?

Was the original design flawed? Why was the original design acceptable before the crash, but not acceptable after?

Was redesign of the tanks a prudent move or an overreaction to the crash?

15) If you flip a coin 10 times, which of the following sequences is more likely to occur?

a) HTTHHHTTHT or b) HHHHHHHHHH

Which of these is more likely?

a) HTTHHHTTHT or b) HHHHHTTTTT

TABLE OF CONTENTS

Preface

1: Introduction

2: Belief Generation and Sensemaking

Chapter 2 is a discussion of how we collect and evaluate data to generate conclusions/beliefs and how we use our beliefs to make sense of data.

3: Objectives and Alternatives

We are goal driven and results oriented. Management values action. We tend to identify alternatives early in the process and to rush off to investigate and/or implement the first reasonable alternative we think of. This leads often to solving the wrong problem for the wrong reasons. A structured focus on objectives improves decisions and designs.

4: Deciding

This is not a very important chapter. Most people have little trouble selecting from among the available alternatives for well-developed problems. This chapter includes: some statistics, a short discussion of Bayesian reasoning, decision methods such as decision trees, optimization methods, utility theory, prospect theory, regret theory, an introduction to decision strategies, and a discussion of the role of emotions in decision making.

5: Acting

This is the most important chapter in the course. Good decisions are not our aim; effective actions are. And many 'good' decisions do not result in effective actions. This chapter investigates why decisions often go wrong in the implementation. Includes important insights into interface management.

6: Learning

An evaluation of why certain types of lessons are hard to learn and why we tend repeat some mistakes and/or learn the wrong lessons from failures.

7: Naturalistic Decision Making – Impacts of Stress and Expertise

Most decision making research is conducted in university labs with college students as subjects. This research is the basis for much of what is contained in Chapters 2 thru 6. Such research often does not translate well to the real world. This chapter investigates the surprising impacts of stress and expertise on decision making.

8: Teams and Interfaces

Covers the use of decision theory insights to improve the performance of teams and to minimize problems at the interfaces between teams.

9: Summary and Suggestions

How to use the learnings and insights from the first eight chapters to make better decisions, do better design work and structure better design organizations. Includes a structured strategy for making decisions.

Appendices
Sources and Recommended Reading
Index

PREFACE

The Challenger Launch Decision – Learning the Wrong Lessons

On Jan 28, 1986 the Shuttle Challenger crashed on takeoff. Several investigations including the very high profile Presidential Commission and a Congressional investigation reached similar conclusions; that the crash was caused by:

- Production pressures on NASA which caused them to ignore risks.
- A flawed safety program with inadequate reporting rules.
- Violation of risk reporting rules.
- Failure of engineers to act decisively to solve a known problem.

NASA took these criticisms to heart and made significant changes in its safety program.

In chapter 2 we'll review the extensive investigation of the accident conducted by Diane Vaughan (15). Vaughan concluded, correctly in my opinion, that all four errors noted above are factually incorrect. What did NASA "learn" from the Challenger crash? A decade later the Kraft Report (1995) noted:

> "As a result of the Challenger incident a "safety shield" philosophy has evolved creating a difficult management situation. Managers, engineers, and business people are reluctant to make decisions that involve risk because of the fear of persecution."
>
> and "…the <safety program> used today may make the vehicle less safe…."

Why did the investigators and NASA learn the wrong lessons? How can we avoid learning the wrong lessons on our projects?

Making Sense and Making Decisions

I am a chemical engineer. I've spent most of my 30+ year career in process and project design roles. It occurred to me fairly early in my career that decision making on projects is both hard and important; I've been interested in the subject ever since.

In the early years I studied mathematical tools used for decision making: linear programming, integer programming, decision trees, multi-attribute utility theory, probability, stochastic programming (Monte Carlo simulation), etc. Those were all fun topics and I used those tools to good effect at times. But those techniques weren't applicable to very many problems. In the cases where I applied them they didn't always yield valuable insight. Worse, people tend to ignore the results of these tools if the results don't support their preconceived opinions or objectives.

It eventually dawned on me that I was studying the wrong things. Most of the hard decisions we face in designing projects are socially driven and constrained and are not readily amenable to mathematical analysis. The difficult decisions frequently stem from differences of opinion between stakeholders. And as the projects I worked on became larger and more complex the social aspects of engineering decision making became more dominant and more difficult to understand and manage.

An engineering project organization is one of the most interesting and complex environments in the world for the study and practice of decision making. Large project organizations are typically composed of several teams working concurrently on different areas of the design. And in the end all the pieces have to fit together in a working whole. Decisions made by each team over the course of a long project have to be consistent with the overall objectives. The usual categorization of decisions as either individual or group decisions doesn't adequately cover the range of engineering project decisions. Some of the decisions are made by individuals, some are made by groups (teams) and some are made by groups of groups.

As we'll see, the errors made in the decision to launch the Shuttle Challenger were caused largely by social constraints imposed by the organizational structure. The errors made by the commissions that investigated the crash are also understandable errors. They discovered the obvious errors; errors that "made sense".

Understanding and dealing with the social nature of engineering decision making is a crucially important factor in successful project design.

The good news is that there is a large body of research available on this subject. But there is more bad news than good:

- The literature base is not simply large; it is absolutely huge. It is far too voluminous and diverse to be approachable by the typical engineer.
- It is not published in books or periodicals that engineers read. This material is found largely in psychology, sociology, cognitive science, mathematics and business journals.
- It is not consistent. In the behavioral sciences different researchers often disagree.

The typical engineer would not know where to start such a study. And unless you read fairly diligently for months, if not years, you will get an incomplete picture, at best. After years of obsessive reading on this subject and years of trying various ideas in practice, I believe that I have something important to share with you about the nature of decision making in engineering practice.

Making Sense and Making Decisions is:

1) A literature survey. A careers worth of reading on the subject of decision making, highly filtered and condensed down into a manageable amount of information.

2) Suggestions for using the findings of a few researchers to improve your decision making skills.

3) Suggestions for using decision theory to improve engineering design processes and engineering design organizations.

Improving the Engineering Design Process

A study of decision theory yields valuable insights for improving engineering design processes. In this course we will seek to generate answers to four questions particularly important to successful project management:

- ***Why do so many changes occur late in projects?***
- ***Why do we have so many problems at interfaces?***
- ***Why do we repeat past mistakes?***
- ***Why do we so often finish behind schedule?***

Not a Course in Psychology

In case you are uncomfortable with this talk of a social basis for decision making, let me assure you that this is not a course in psychology. We do not have to investigate deep-seated disorders in the human psyche to understand human behavior on project teams. We will not investigate how different people with different personalities behave differently. We are all much more alike than we are different. The problem with engineering teams is that everyone behaves the same; everyone behaves the way you do. Understand your own behavior and you will understand the behavior of others.

Seminar Course

This book was written as the handout for a seminar course and still serves that function though it can also be read independent of the course.

This book/course will make you a better engineer.

Howard Duhon, P.E.
Systems Engineering Manager
GATE
Houston, TX
www.gateinc.com
hduhon@gateinc.com

Chapter 1
Introduction

Sources:

In developing this book I have used ideas from numerous researchers. The most important ideas come from:

Chris Argyris: ....Action Science
Ralph Keeney: ....Value-focused Thinking
Gary Klein: ....Naturalistic Decision Making
Karl Weick: ....Sensemaking
Amos Tversky, Daniel Kahneman: ......Biases and Heuristics
Gerd Gigerenzer: ....Decision Strategies
James Reason: ....Human Error Research

Preface:

Research is usually focused study on a small aspect of a larger field of inquiry. Each of the researchers noted above and others referenced in the course has studied an aspect of the field of decision making. In their areas of expertise, each has made a fundamentally important contribution. But because their work is focused, it is limited in applicability, and in some cases I believe fundamentally flawed when judged from a holistic point of view. What makes this course unique is the organization of ideas selected from numerous individual researchers into a comprehensive coverage of the subject of decision making.

The organization of the course is the main topic of Chapter 1.

If the only tool you have is a hammer,
everything looks like a nail.

CHAPTER 1 CONTENTS:

1.1 DECISION MAKING IN ENGINEERING
1.2 CONSCIOUSNESS AND DECISION MAKING
1.3 ALL DECISIONS ARE MADE SUBCONSIOUSLY!?
1.4 EMOTIONS IN DECISION MAKING
1.5 PARADOX – LESS IS MORE
1.6 OBJECTIVES OF THE COURSE
1.7 ORGANIZATION OF THE STUDY OF DECISION MAKING
1.8 ORGANIZATION OF THE COURSE
1.9 WARNING – MOST BUSINESS ADVICE IS BAD ADVICE!

1.1 DECISION MAKING IN ENGINEERING

The Nature of Design Decision Making

When we think of decision making we tend to think of large decisions such as where to build a new plant and/or discrete decisions such as whether to build a vessel out of carbon steel or a corrosion resistant alloy. But there is much more to decision making in project work. Consider the typical progression of an engineering task as illustrated in the simplified engineering model below:

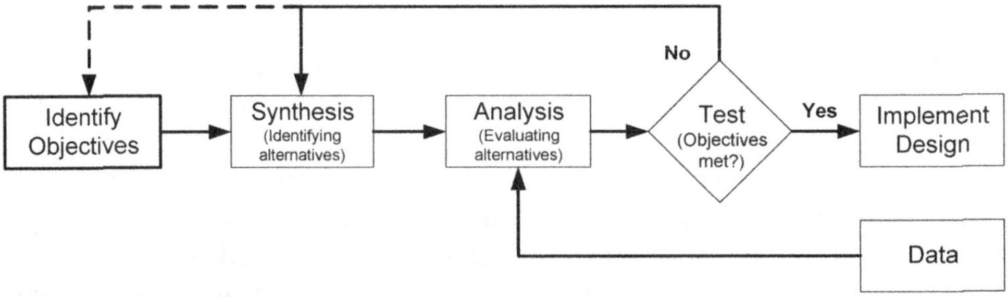

- The model begins with identification of **objectives**.
- Having identified objectives we attempt to identify design alternatives which will satisfy those objectives (**synthesis**).
- A promising design option is evaluated against the design objectives (**analysis**).
- We then ask whether the design option satisfies all of our design objectives (**test**). If the design option satisfies all of the identified objectives we may proceed to **implement** the design option (detailed design, construction, etc.).
- On the other hand, if the design option under consideration does not satisfy one or more of the identified objectives we return to synthesis and try to identify a better alternative.
- The dotted line suggests that our objectives change over the course of the design. In the course of the design effort we naturally identify some objectives we initially missed. And if we fail to identify an alternative which satisfies all of our objectives, then we will eventually be forced to relax one or more of our objectives.

This process has one decision point; deciding whether the alternative you are analyzing is acceptable. But there are many other decisions in this process including:

- The objectives are not givens. We <u>decide</u> which objectives are valid and which are most important.
- We <u>decide</u> which alternatives to consider seriously, which to ignore, and when to stop looking for a better one.
- We <u>decide</u> how to evaluate the promising alternatives including what data to collect and what analysis methods to use.
- We <u>decide</u> whom to consult when we have problems and whose opinions are valuable.

The success of our projects depends on how well and how quickly we make all of these decisions.

The Number of Decisions we make

How many decisions does it take to execute a project? Many more than you might imagine. Consider the installation of a simple run of pipe. Dozens of decisions have to be made. The table below is a partial list. Many of these decisions are trivial for experienced engineers and many are made easier by the existence of codes and standards. But they all need to be made and many of these simple decisions will "bite" you if you get them wrong. If this simple task requires so many decisions consider how many millions of decisions are needed to execute a large project.

Decisions Required to Design and Install Pipe (partial list)	
Design code (B313, B31.8, API RP 111, etc.) Pipe diameter (flowrate, fluid composition range, physical properties, pressure drop, velocity, erosion, strength (support req.)) Material of Construction (internal corrosion, external corrosion, strength, design life (design for life of facility or expect replacement) Pressure rating (normal pressure, incidental pressure, code allowances) Number, type and location of valves Pipe thickness (pressure rating, corrosion allowance, support strength, commercially available thicknesses, material of construction) Welding spec/procedure, weld test procedure Connection type(s) [welded (butt weld, socket weld, other), flanged, screwed, specialty connection] Flanged joint specification (raised face, ring joint, specialty), Bolt specification (type, material of construction, coating) Bolted joint integrity management plan including flange makeup procedure	Joint Leak Test Procedure Number and location of connections (minimize to avoid leaks, maximize to simplify maintenance, tie-ins) Pipe routing (pipe rack or direct, avoid low spots? minimize length? minimize bends? accessible for operation, maintenance, testing) Bend and tee radius and type (for pigging, erosion) External coating (type, thickness, application method, color (aesthetics or color coding?)) Internal coating? Cathodic protection? (if underwater) Line supports – type, spacing Managing pipe stress as caused by temperature change (loops, stress relief, spring supports) Location of vents and drains (for maintenance, hydrotesting) Hydrotest plan Over-pressure protection Insulation (heat conservation of personnel protection) Heat tracing? Fire protection

1.2 CONSCIOUSNESS AND DECISION MAKING

The discussion of consciousness in this section and the next is based on the excellent book *Origins of Consciousness* by Julian Jaynes (19).

You are not consciously aware of all the decisions you make. Have you ever been challenged to defend a decision that seemed like a no-brainer, only to find that you were unable to come up with convincing arguments? Our brains are capable of making decisions based on the flimsiest of evidence. These decisions often seem obvious to us until challenged by others. A surprising amount of our mental activity occurs outside of our conscious thought. It is instructive to examine the relationship between consciousness and decision making.

It is difficult to define what consciousness is so I begin with a short discussion what consciousness is not:

Memory is not a Conscious Activity

You meet someone you know and immediately call them by name without "searching" your memory. But sometimes you cannot come up with the name of a person you know. Thinking hard about their name rarely works. Forget about the effort for a few seconds though, and the name often pops up as if by magic. Obtaining a person's name from your memory is not normally a conscious activity, and consciously searching your memory is often not very effective.

Learning Occurs Subconsciously

Learning does not depend on consciousness. Pavlov's dogs learned to salivate on hearing a bell.

We learned our native languages as children without any conscious intent. In contrast, learning a second language is usually a conscious effort and is much harder to accomplish.

Problem Solving Occurs Subconsciously

Scientists refer to the three B's of creativity; bed, bath and bus. Out-of-the-box ideas frequently occur when our minds are completely off the subject (such as when we are taking a bath or riding a bus). Kekule said that he actually came up with the structure of the benzene molecule in a dream.

Some Decisions are made Subconsciously

When you are preoccupied with other thoughts while driving, you will usually stop at a red light and drive through a green light without being conscious of the decision.

A piano player sees a note on sheet music and reaches with the correct finger to play the correct note and may do so several times in one second, often playing multiple notes simultaneously. Each decision is outside of consciousness. Indeed, if she becomes conscious of the notes she is seeing and the keys she has to hit, she will lose the ability to play.

The same phenomenon occurs in typewriting. Hitting the keys is automatic. If you have to think about the keys your speed decreases to a crawl.

It should be clear that some decisions are made subconsciously and that many of them must be made subconsciously. The fact is that all decisions are made subconsciously.

1.3 ALL DECISIONS ARE MADE SUBCONSIOUSLY!?

I know you are protesting this. In the previous section we demonstrated that we make many decisions subconsciously even when preoccupied by other thoughts, such as playing piano keys and stopping at a red light. Clearly some decisions are made subconsciously. It is a leap to argue that all decisions are subconscious, but that is exactly what I am claiming.

For difficult decisions the data gathering, the deliberations, the comparing of alternatives are all, or mostly, done consciously. But then you make a judgment, form a conclusion – where did that come from?

Try to focus on the moment in time in which you go from undecided to decided. Describe the process. How does it work? What spark causes you to make the leap to a decision? You cannot do any of these. The actual phenomena of making a decision is outside of conscious thought – whether that decision is which key on the piano to play with which finger or whether to use chemotherapy or surgery to fight a cancer. You cannot consciously flip the switch that takes you from undecided to decided. That switch is outside of conscious thought and control. Selection between options apparently occurs in an old part (evolutionarily speaking) of our brain that is not open to conscious introspection.

So, if decision making is subconscious can we improve our decision-making skills? Yes. Just as it is possible to improve a golf swing (a largely unconscious activity), study and practice will improve your decision-making skills.

Model of Cognition

It's not just decision making that happens subconsciously. Reason (28) describes the brain as composed of two systems, the automatic subsystem and the attentional subsystem. (Some authors refer to these as 'System 1' and 'System 2' respectively.)

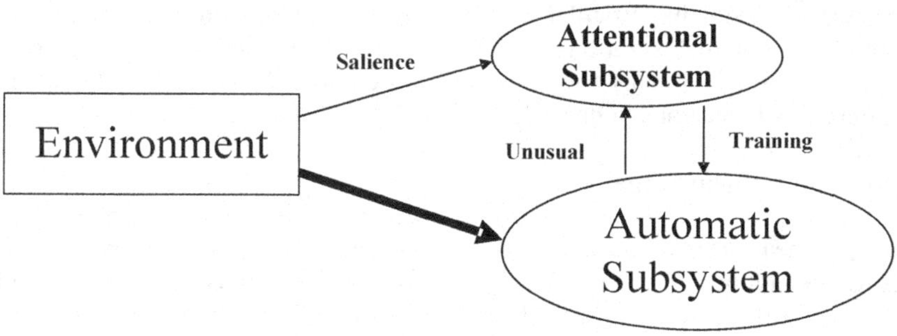

Automatic Subsystem: The skills that we take for granted (walking, bicycle riding, driving a car, speech, recognizing faces, etc.) are the result of rules (schemas) stored within the automatic subsystem. The automatic subsystem acts below the level of conscious attention. It is a massive parallel processor. This system stores a virtually unlimited number of schemas to be applied in the appropriate situations. Schemas are selected on the basis of two mechanisms, similarity matching and frequency gambling.

Similarity Matching (pattern recognition): We detect patterns without being conscious of the detection. We recognize faces, meaning in speech, etc. When a situation sufficiently matches a stored pattern, then the schemas relevant to that situation are activated.

Frequency Gambling: Data from the environment is often ambiguous. We are not always sure of what is going on. In such situations the automatic subsystem tends to select schemas that have worked in the past in similar, if not identical, situations. The frequency of previous successful use of a particular schema makes it more likely that it will be applied in an ambiguous situation.

Attentional Subsystem: Despite all this brainpower at work, we are conscious of only a limited amount of data at one time. The attention of the attentional subsystem is grabbed by salient features in the environment and by selected results of pattern matching within the automatic subsystem. The attentional subsystem develops and applies rules until they are "learned" by the automatic subsystem. For instance: learning to play the piano is a conscious activity until the automatic subsystem learns the rules.

The automatic subsystem is a huge and powerful parallel processor capable of many simultaneous operations. But our conscious, or attentional subsystem, is a serial (one thing at a time), slow, logic processor with very little memory (short term memory). It is difficult to maintain attention for long periods.

1.4 EMOTIONS IN DECISION MAKING

Rene Descartes was in a Paris bar late one night as closing time approached. The barkeeper announced "last call" and asked Rene if he wanted "one for the road?" Rene responded "I think not" and pfft, he disappeared.

Joke – author unknown

It's one of my favorite jokes. Descartes' line **"I think, therefore I am"** is one of the most famous sayings of all time. It captures our culture's philosophy on reason and emotion.

In business decisions in particular, a common view is that cold calculated reason should be employed and that emotion has little place. Damasio (11) reported on the very interesting story of Phineas Cage, a story which forces one to a different conclusion about the role of emotions in decision making.

The Story of Phineas Gage

In 1848, Phineas Gage, a railroad foreman, was tamping gunpowder into a hole. It was a standard operation. But this time a momentary lack of concentration resulted in a missed step in the process. An inadvertent spark set off an explosion. The tamping rod, 1 ¼" in diameter and 3'7" long, was sent flying through Gage's head. Gage did not die; never even lost consciousness, and by all outward appearances made a complete recovery and returned to work. But the Phineas Gage who returned to work was a changed man. Friends and associates reported that he was "a different person". He was unable or unwilling to perform his duties and was soon fired. He lived the rest of his life wandering from job to job.

Gage was not mentally impaired in any usual concept of the word. His memory was intact. His reasoning was rational. He communicated effectively. He appeared competent and in most aspects unchanged from the pre-accident Gage. But there was a dramatic change. **The post-accident Phineas Gage had a much diminished capacity to feel emotions.** Scientists have since identified areas of the brain which govern our feeling of emotion. One of these areas was damaged in Gage's brain. Mr. Gage had another disability as a result of the injury. He was largely unable to make decisions in social settings.

We make decisions because we care. Emotions are the root of caring. Phineas Gage could not decide, because he didn't care. He didn't care because he did not feel emotions.

1.5 PARADOX – LESS IS MORE

We may decide where to eat lunch by flipping a coin. Or we might choose a restaurant on the basis of the one that serves the best roast beef sandwich. Simple problems suggest simple solution strategies.

But if we have to decide on the best routing of a new highway through sensitive wetlands, we may spend years collecting and evaluating data, obtaining the opinions of interested parties, weighing pros and cons, and ultimately trying to sell our opinions to others. Difficult decision problems suggest the need for complex analytical decision strategies.

We tend to use simple decision strategies for simple problems when little is at stake and to use more analytical strategies for complex problems when a great deal is at stake.

It has been shown, in some remarkable research, that simple decision strategies, such as choosing the alternative that best satisfies our most important objective, are often superior to very sophisticated techniques. In *Simple Heuristics That Make Us Smart* (14) Gigerenzer makes a strong case for keeping it simple.

Heart Attack Patient Risk Assessment

To illustrate the point, Gigerenzer reports on the technique used by San Diego Medical Center (SDMC) for classifying heart attack patients prior to discharge as high-risk or low-risk. Clearly this is a complicated environment. SDMC personnel measure and record up to 19 risk factors on each patient. Common sense dictates that all available data should be used to assess patient risk. Indeed, programs have been written to assess risk based on statistical regression of all 19 factors. But the method used by doctors (illustrated below) is very simple. It uses only three factors at most, and all three are yes-or-no questions.

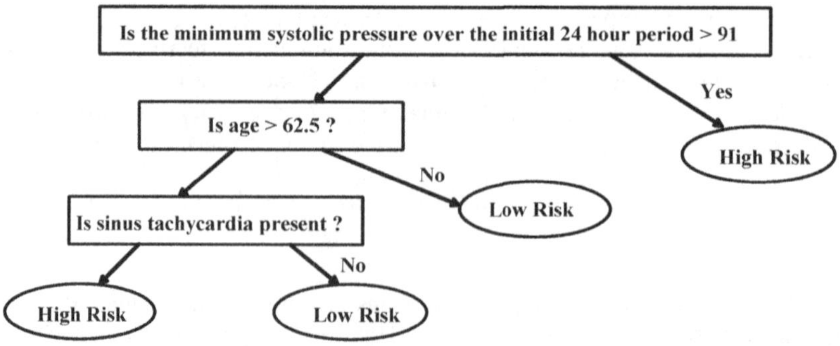

This may seem like a shortcut and an invitation to medical malpractice suits, but it is not. *This simple method is used because it actually performs better than models based on statistical regression of all available data.*

Using too much data can actually degrade performance. The explanation of this result is:

1) regression of large amounts of data tends to yield, by happenstance, correlations that are true only of the sample population, not of the population as a whole, and
2) when collecting large amounts of data, it is likely that some of it will be incorrect.

KISS – Keep It Simple Stupid

Complicated decision strategies only make sense when

a) the environment is complicated,
b) our model of the environment accurately reflects the complexity of the environment, and
c) we have enough data, and accurate enough data, to effectively use the model.

Most environments we deal with are actually fairly simple. Even for complex problems, a few variables often dominate. In those cases, accurate assessment of the important information is often more important than a comprehensive assessment of the problem.

The adage KISS (keep it simple stupid) tends to be good advice.

1.6 OBJECTIVES OF THE COURSE

I hope that the last few sessions have piqued your interest and that I can maintain your interest throughout the book. The points I've tried to make in the sections above include:

- We make a tremendous number of decisions when we execute a project.
- We are not consciously aware of making many of them.
- The methods we use in making decision have an impact on the quality of those decisions.
- There are so many decisions to be made that formal decision methods cannot be the answer for most of them.

What is the answer? How do we improve decision making on our projects? I think that the answer is to develop a good fundamental understanding of decision making and to apply this understanding to the design process. That is the objective of this book.

Literature Review

Chapters 2 through 8 provide a literature review of interesting and useful findings by researchers in decision theory.

Improving your decision making skills

In Chapter 9 I bring together insights from the first 8 chapters to develop a decision strategy applicable to any type of decision problem.

Improving the Engineering Design Process

In addition to improving your decision making skills I will attempt to show you how we can use insights from decision theory to improve engineering design processes. In particular we will seek to generate comprehensive answers to the following questions:

- ***Why do so many changes occur late in projects?***
- ***Why do we have so many problems at interfaces?***
- ***Why do we repeat past mistakes?***
- ***Why do we so often finish behind schedule?***

At the end of each chapter we'll ask these four questions and try to determine whether anything we've learned in that chapter identifies causes for those problems.

Would you like to answer a different question?

1.7 ORGANIZATION OF THE STUDY OF DECISION MAKING

There is a great deal of information in this book, and figuring out a way to organize it was challenging. There are many possible ways to organize a study of decision making including:

- By the decision making technique used.
- By the number of people making the decision (individual or group).
- For group decisions, by the method and level of aggregation of individual knowledge.
- Zero-sum games (win-lose) vs. Non-zero-sum games (potential for win-win).
- Analytical vs. Naturalistic.

None of these approaches works for the type of discussion I want to encourage.

Normative, Descriptive, Methodological

A great deal of decision making literature is organized around the concepts of normative vs. descriptive models of decision making. Although the course is not organized around those concepts, this topology had great influence on the development of the material and is worth discussing here.

<u>Normative</u>

Normative methods refer to <u>the way thing should be done</u>; the way to determine logically the best answer. This is clearly an appropriate approach if all the variables and objectives can be defined numerically and unambiguously. Examples include linear programming used to determine the optimum way to run a factory. Mathematicians have extended this to a definition of rationality as it applies to preference among choices.

<u>Descriptive</u>

But people do not always behave rationally. Psychologists and sociologists get involved in the game when it comes to <u>explaining how people actually behave</u> as compared to what we describe as rational.

<u>Methodological</u>

Decision analysts attempt to develop methods to solve difficult decision problems. While this course is not about such techniques, we will discuss and practice some of the more commonly used ones including decision trees and multi-attribute utility theory. As we will see, such techniques should be used very carefully. They are designed to capture our values with sufficient accuracy that the decision can then be "calculated". That is very difficult to do.

1.8 ORGANIZATION OF THE COURSE

So, after that long-winded discussion of how the course is not organized, here is how it is organized. The sketch below is an influence diagram of the decision process.

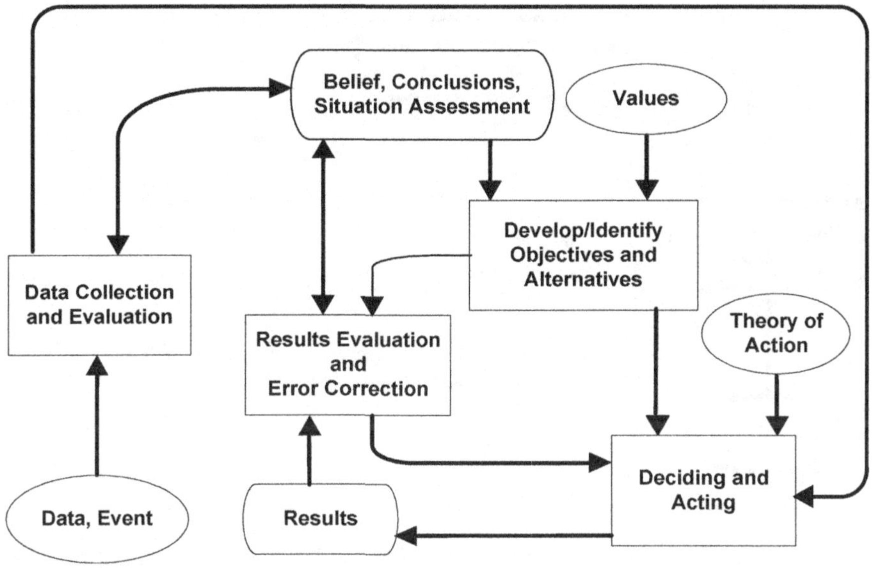

Stepping through this diagram (begin at lower left):

1) We begin with available data or events.
2) We "collect" or "see" some of that data, often a triggering event, and evaluate the data in some way
3) to generate conclusions, beliefs, or situation assessments.
4) If we care (values) about the conclusions we've come to
5) we may develop objectives to change things. If we can identify alternatives that may achieve those objectives
6) we will evaluate those alternatives, decide which is best, and take some action to implement it. The action we take is influenced by our theory of action (chapter 5).
7) Results of actions taken may or may not match our expectations. If there is a mismatch (error) we may take action to correct the error (learning). This learning can be simple correction of an implementation flaw, or it can be a more difficult reassessment of fundamental assumptions and beliefs.

I've divided this influence diagram into 4 activities for study:

- Generating Beliefs and Making Sense.
- Identifying Objectives and Alternatives.
- Deciding and Acting.
- Learning.

These topics form chapters 2 thru 6 of the course as discussed briefly below.

Generating Beliefs and Making Sense (Chapter 2)

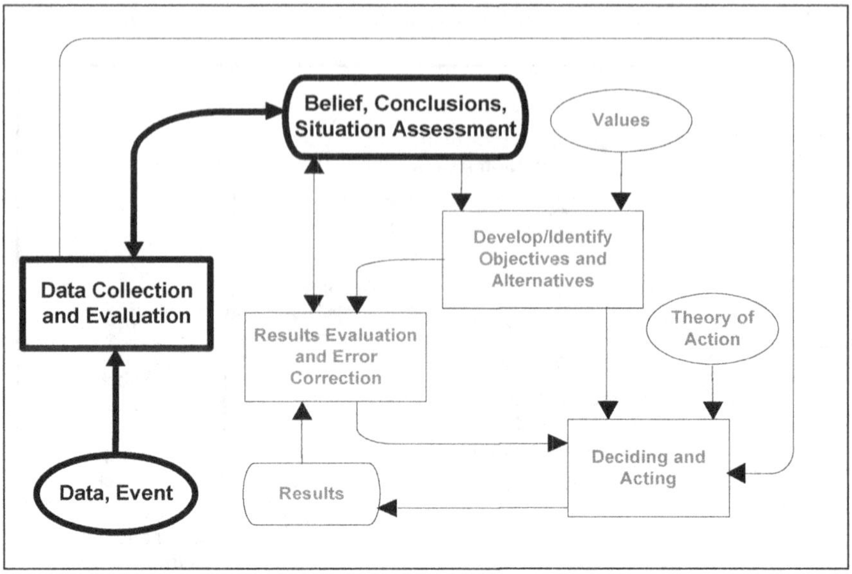

Data exists. We "collect" or "see" some of that data, and evaluate the data in some way. Evaluation of data generates beliefs, conclusions, or assessment of the situation. The double edged arrow between the "Data Collection" node and the "Belief .." node is important.

The data we collect affects our beliefs, and our beliefs affect the data we collect and how we evaluate the data.

Chapter 2 is a discussion of how we collect and evaluate data to generate conclusion/beliefs and how we use our beliefs to make sense of data.

Identifying Objectives and Alternatives (Chapter 3)

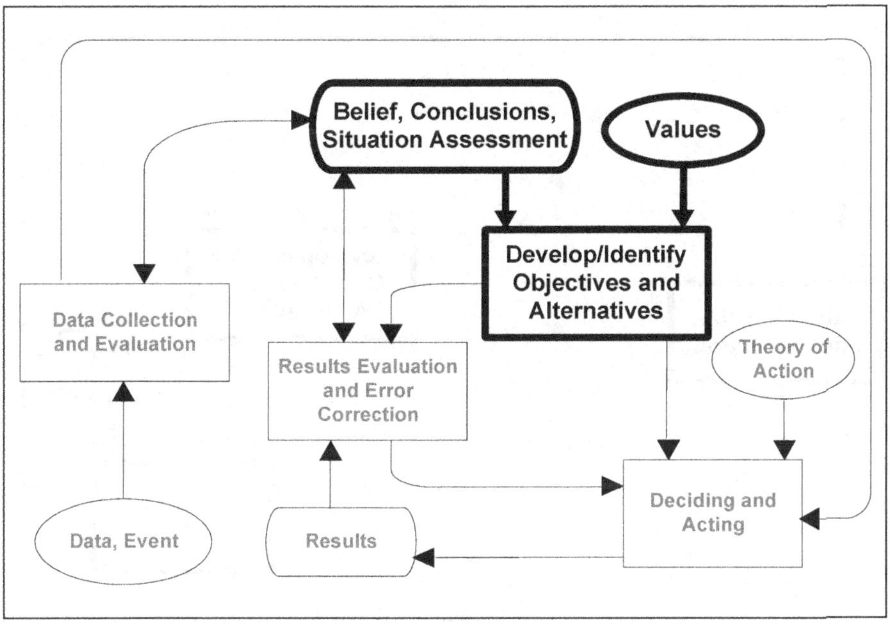

If you **care** about the things you've come to believe, you may develop objectives to change things and try to identify alternatives which might achieve those objectives.

Caring is based on your **values**.

You make decision because you care. Your values are the reasons you care.

Chapter 3 includes a structured approach to using your values to better identify your objectives and alternatives. The chapter is based largely on the work of Ralph Keeney (2).

Deciding and Acting (Chapters 4 and 5)

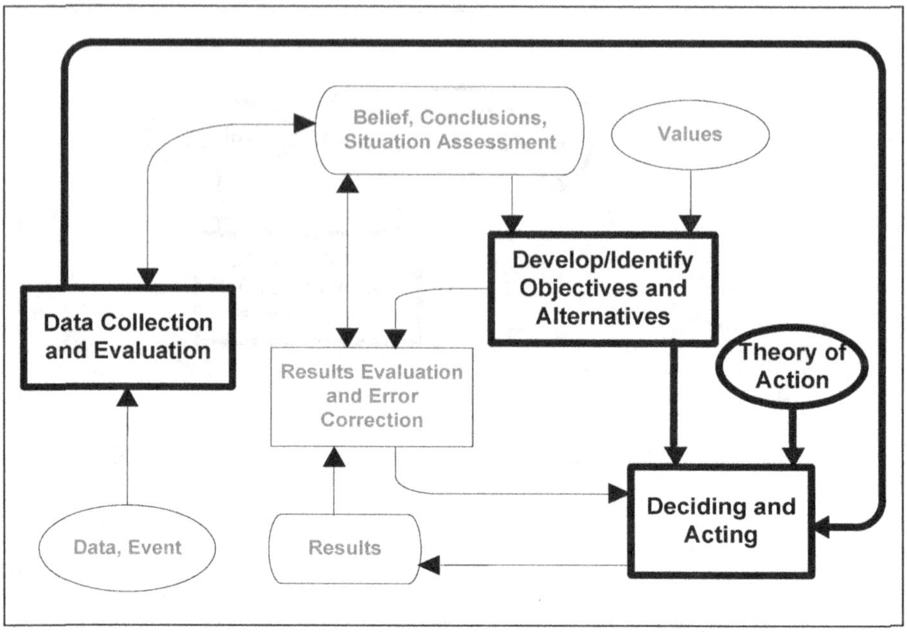

Chapter 4 (Deciding) focuses on methods of selecting the best alternative, including mathematical methods and decision strategies and preference theory models.

Chapter 5 (Acting) focuses on action. **Chapter 5 is the most important chapter in the course.** In particular, the concept of **theories of action** is the most important concept in the course. Good decisions are not our aim; effective actions are. And many decisions do not result in effective actions. In this chapter we investigate why "good" decisions often go wrong in the implementation.

Learning (Chapter 6)

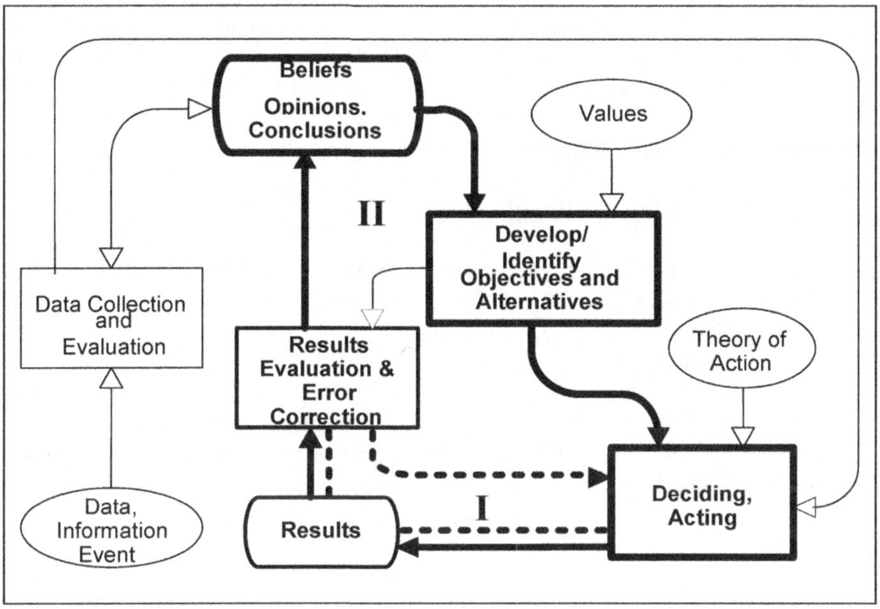

Why is learning hard? Why do we repeat mistakes?

> The chapter focuses on why certain types of lessons are difficult to learn and why we tend repeat some mistakes and/or learn the wrong lessons from some failures.

Two learning loops are explored. The learning loop in dashed lines labeled 'I" is easy, non-threatening learning. The loop in solid bold lines labeled 'II' requires reassessment of beliefs and is much harder. The more important a lesson is; the more difficult it is to learn.

Impact of Stress and Expertise (Chapter 7)

The literature on decision making is remarkably large. An amazing aspect of all this literature is that, until very recently, essentially none of it considered how people actually make decisions in the real world!

Most decision–making research is conducted by university professors using students as subjects. In 1985 Gary Klein conducted a study of decision making by firefighters in emergency situations (3). His observations led to a new field of study called Naturalistic Decision Making (NDM). Researchers in this field now study how real people in the real world make decisions in settings in which they have some expertise. Often these situations involve time pressure or stress. Results of this research have revolutionized our understanding of decision making under stress.

Teams and Interfaces (Chapter 8)

Teams are complex and misunderstood entities. Stories about miracle working "skunk teams" create unrealistic expectations of teams. Most teams have problems that negatively impact their performance. The aim of chapter 8 is to apply sound theory on team processes to the task of improving the performance of teams.

Summary and Suggestions (Chapter 9)

<u>Causes and Solutions of Engineering Problems:</u>

Chapter 9 summarizes the causes of engineering problems identified in chapters 2 thru 8 and identifies possible solutions.

<u>General Decision Strategy</u>

One of the observations of the course is that we often apply ineffective decision strategies. In Chapter 9 I suggest a general strategy for approaching decision problems. The strategy is based on key observations from the course.

1.9 WARNING – MOST BUSINESS ADVICE IS BAD ADVICE!

A warning is appropriate here. This course provides business advice. <u>Most business advice is bad advice!</u>

Consider how often the business advice du jour changes. Bookstores shelves are crowded with books touting the latest business advice. Each new idea seems to work for a while and then fails and disappears or at least loses followers. Five years from now few of the books currently on the shelf will still be for sale. A new crop suggesting different solutions will take their place. Obviously, most business advice is bad advice without staying power.

Since most business advice is bad advice
and this course gives business advice
then the probability is high that this course provides bad advice.

What makes me think that I can beat the odds and provide you with effective advice that has staying power?

I believe the reasons that most business advice is bad advice are well understood by practitioners of action science. This course will give you tools to judge whether the advice you receive is good advice or not (chapter 5). I encourage you to apply those tools and judge for yourself whether the advice provided in this course is likely to be effective.

Chapter 2

Belief Generation and Sensemaking

Sources:

Amos Tversky and Daniel Kahneman (4, P1, P2) identified numerous biases that we fall prey to in evaluation of data.

In *Sensemaking in Organizations, (*5) Karl Weick provides a summary of the field of sensemaking.

Diane Vaughan applies many of the insights of the field of sensemaking in her analysis of the space shuttle Challenger disaster in *The Challenger Launch Decision* (15).

Preface:

Our decisions begin with, and are heavily influenced by, our beliefs. Our beliefs are determined by the information we are exposed to and the way we evaluate that information. The generation of beliefs is often a haphazard process. This chapter covers important features of our belief generation process and identifies sources of error.

Seeing is <u>not</u> believing, sometimes we must believe in order to be able to see.

CHAPTER 2 CONTENTS

2.1 INTRODUCTION TO BELIEF GENERATION
2.2 REASONS FOR BELIEF
2.3 DATA EVALUATION HUERISTICS
2.4 REPRESENTATIVENESS
2.5 AVAILABILITY
2.6 REFERENCE EFFECT
2.7 EXERCISE: SETTING NEGOTIATING LIMITS
2.8 EXERCISE RESULTS
2.9 ANCHORING
2.10 RISK PERCEPTION AND THE AFFECT HEURISTIC
2.11 COPING WITH EMOTIONS
2.12 SOCIAL PRESSURE
2.13 ON SELECTING AND EVALUATING DATA
2.14 ASSIGNING MEANING – CREATING INFORMATION
2.15 SENSEMAKING
2.16 PROPERTIES OF SENSEMAKING
2.17 DISSONANCE REDUCTION
2.18 SENSEMAKING IN ORGANIZATIONS
2.19 SENSEMAKING IN CHALLENGER LAUNCH DECISION
2.20 GROUP IMPACT – STRUCTURAL SECRECY
2.21 CHAPTER SUMMARY – UNDERSTANDING MISUNDERSTANDINGS
2.22 ANSWERING THE FOUR QUESTIONS

2.1 INTRODUCTION TO BELIEF GENERATION

Our decisions begin with, and are heavily influenced by, our beliefs. But we all frequently believe things that are not true. In this chapter we investigate the processes by which we use data and information to craft our beliefs.

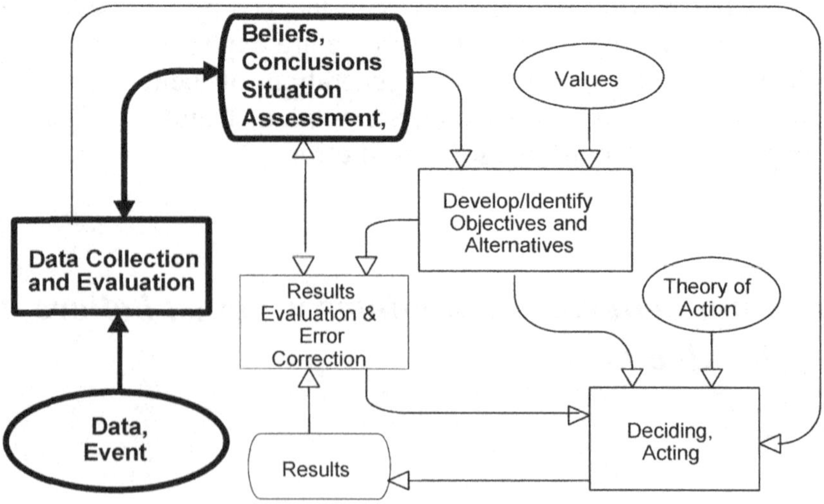

By 'beliefs' I'm referring to all manner of belief from strong beliefs such as your religious faith and your political persuasion to more weakly held beliefs such as your assessment of the cause of a plant upset or your faith in a particular pressure drop estimation method.

The process of belief generation is shown highlighted on the decision process influence diagram above.

- There is a pool of data available.
- We can rarely, if ever, learn all there is to know about a problem. We must choose, by some method, which data to collect or "see".
- We apply some criteria to evaluate the data we have collected.
- The collection and evaluation of data leads us to formulate opinions, conclusion, beliefs about the state of the world.

A very important part of this diagram is the double arrow between the data collection node and the belief node. Data impacts our beliefs, but our beliefs affect which data we choose to collect and affect how we evaluate that data. As Albert Einstein noted, "Our theories determine what we measure." Beliefs tend to be self-reinforcing, whether they are true or false.

There are two main topics in this chapter. First we'll evaluate how we collect and evaluate data in order to generate beliefs. Second we'll evaluate how beliefs impact our collection and evaluation of data.

Generating Beliefs

Data Evaluation Heuristics

We cannot rigorously evaluate all the data we see, partly because most of us lack the mathematical skills/knowledge to do it and partly because the world is passing by too rapidly to allow us the luxury. We therefore apply some simplifying rules, or heuristics, to speed the process of data evaluation.

Coping with Emotions

When an event is difficult to deal with emotionally we sometimes cope by evaluating the event in a less threatening way. In the extreme case we may even deny that it happened at all.

Social Pressure

Social pressure influences us to believe certain things.

Inference

Much of what we take to be facts are actually inferences.

Maintaining Beliefs

Sensemaking

Sensemaking is, in effect, the opposite of belief generation. When we believe something, we tend to make sense of new data as supporting that belief.

2.2 REASONS FOR BELIEF

When my love swears that she is made of truth, I do believe her, though I know she lies ...
Shakespeare, Sonnet 138

Why do you believe the things you believe?

Scientific Evidence:

Because of your scientific training you probably seek to base your beliefs on verifiable evidence. Verifiable scientific evidence is certainly one of the reasons we believe, but there are others.

Social Pressure

We believe some things because it is socially advantageous for us to believe them. The most common example of this is your religious belief. Most people share the religious beliefs of their parents.

If you hang out with a group of way-out leftist liberals and value their friendship, you will tend to believe the things they believe about welfare, the environment, the oppression of the working classes, etc. And your beliefs will likely persist even in the face of scientific studies showing them to be wrong.

Salience

You believe some things just because they are easy to believe. A study of the religious beliefs of people all over the world (21) suggests that all religions feature certain similar beliefs. The apparent explanation for this is that some ideas are easier to believe than others and these salient ideas are therefore more likely to survive and to be repeated and hence, to form the basis for religious thought.

It is claimed that playing classical music to unborn babies makes them smarter. I don't know if there is any scientific basis for this claim; I suspect that there isn't. But it is a claim that is salient; it's easy to believe.

In their runaway best seller *In Search of Excellence,* Tom Peters and Robert Waterhouse list eight traits of excellent firms. They present no scientific evidence for their conclusion (no evidence that excellent firms are different from non-excellent firms in these eight ways). Despite this lack of evidence, millions of people believe their conclusions and have bought the book. The eight traits are salient; they seem obviously true; and we want them to be true, hence we do not require scientific evidence.

Many of our common business beliefs are salient and are hence resistant to refutation. Examples:

- Fixed price bidding of contracts decreases project costs and risks.
- Using more subject matter experts improves designs.

2.3 DATA EVALUATION HUERISTICS

The world passes us by at breath-taking speed. We do not even see, let alone concentrate on, most of the data available to us. The heroes in detective mysteries usually triumph by seeing evidence that was in plain sight, but that everyone else missed. What data we do see we do not have time to process in any rigorous way. Statistical analysis of collected data often shows facts not readily apparent on casual observation, but we do not usually have the time to apply such techniques. And few of us have the mathematical skills to do it even if we wanted to. Yet, we must evaluate data in order to form opinions. The approach we take is to apply simplifying rules, or heuristics, to make the process faster and more available. Many such heuristics have been identified. Several were studied in great detail by Tversky and Kahneman [see (4), (P1) and (P2)]. We'll consider four of them:

Representativeness:

If A is similar to B in some ways, we tend to think it must be similar in other ways as well.

Availability:

Availability is bias due to the easy recall in memory of an example case that may not be representative of other cases or of the probability of occurrence.

Anchoring:

When we make judgments in the face of uncertainty, we often start at some point and then adjust from that point as additional data is available. The starting point (anchor) often has little or no rational basis. Once an anchor is in place, we tend not to move effectively from the anchor point when new data is available.

Reference Effect:

We tend to feel the loss of something we possess more strongly than the failure to gain something of equal value. We overvalue losses and undervalue gains.

Affect Heuristic:

If you have positive feelings towards an activity you will tend to judge the activity to be relatively low in risk and relatively high in value. Conversely, if you have negative feelings towards an activity, then you will tend to judge the activity to be relatively high in risk and relatively low in value.

2.4 REPRESENTATIVENESS

Representativeness errors are a family of errors related to inappropriate correlation of data. If a car has a nice coat of paint we assume it has a good engine. If a car's engine is clean we assume that it will run well. Tall people are assumed to possess better leadership skills.

Insensitivity to Background Probabilities:

If a young tattooed biker applying for an entry level job and a 50 year old well-respected accountant both test positive for marijuana use, we assume they are both guilty. But the test is not perfect and there is a chance of error in both cases. The odds are much higher that the accountant's results represent a false positive. Background probability should affect our interpretation of new data. We will discuss that when we cover Bayes theory in Chapter 4.

Insensitivity to Sample Size:

The smaller the sample size, the more likely it is that the results are not representative of the truth, yet our tendency is to believe we know the answer even if we have only one data point. This bias is particularly prevalent when we try something new and it works the first time. This one success sells us on the method and it may take many future failures to convince us otherwise.

Misconception of Chance / Misconception of Regression:

One example of this is the gambler's fallacy. If a coin has come up tails many times in a row, people tend to believe that the odds of it coming up heads on the next flip are higher than 50/50.

Unscrupulous or statistically illiterate entrepreneurs sell programs to help you win the lottery. Oddly enough, these are often based on the opposite belief; that if a number has come up several times lately, it is "hot" and will continue to come up often in the near future.

The law of averages says that if you flip a coin 100 times, on average it will come up heads 50 times. If the first 10 flips come up heads we tend to believe the coin will try to catch up later; maybe 50 of the next 90 flips will be tails so that at 100 we'll be 50/50. But the coin doesn't know it's behind and can't "try to catch up".

The Halo Effect

If you meet someone at a party and you like him because he is very funny you will tend to think that he will be a good employee. You have no good evidence of this; all you know is that he is funny. Since being funny is good, you will tend to think that he is a good person in other ways as well.

Illusion of Validity – Confidence Bias

We often tend to have more confidence in the results of a process than past results justify. Interviews to select among job applicants are notoriously poor predictors of job performance, yet we believe pretty strongly in the process.

Engineers calculating results with a familiar formula tend to view the results as truth. In fact, the formula is usually an empirical fit of experimental data with a significant potential error. We tend to develop trust in techniques we have used in the past, whether or not we've had an opportunity to verify the results. Because we are familiar with the technique, we assume validity of the answers.

Example: Amos Tversky told a story about an Israeli Air Force officer who claimed that criticism was more effective motivation than praise. The officer noted that when he praised pilots after very successful practice exercises, they usually did more poorly the next time out; and when he chastised pilots after very poor exercises, they usually did better the next time out. This officer did not understand regression to the mean. On any given day it is most likely that a pilot's performance will be about average. By simple chance it is likely that a pilot who did very poorly one day would do better the next, and that a pilot who did very well one day would do worse the next day. The officer also suffered from the illusion of validity of his interventions. It is possible that his interventions had little impact.

2.5 AVAILABILITY

One of the most persistent forms of bias is availability error. Availability error occurs when some thoughts are more available in memory, or are more imaginable.

Biases Due to Retrieveability of Instances from Memory:

Early in my career a crane working in a nearby operating unit dropped an object during a construction project. I was a bit shocked by that. A few months later a crane working in my unit dropped a valve. Crane drops happen, or course, but they are pretty rare. For years I overestimated the risk of dropped objects, and overworked the issue on my projects because it was so easy to "see" it happening.

Investigators have concluded that the Concord crash was caused by a piece of steel left on the runway. The metal object or pieces of the tire were projected into a fuel tank causing the crash. The designers of the Concorde knew that foreign objects on the runway posed a threat; hence there was a procedure to sweep the runway prior to each takeoff. The runway was not swept prior to the ill-fated takeoff. Hence, the cause of the crash was:

- Failure to sweep the runway,
- Resulting in projection of object(s) into the fuel tank.

The response to the crash by British Airways and Air France was to ground the planes for months while the fuel tank was modified to make it more resistant to flying debris.

If the result had been less catastrophic the remedy would, no doubt, have been to more diligently enforce the procedure to sweep the runway. The procedure to sweep the runway was deemed adequate protection up to that point. If it had been swept that day, the crash may not have happened. If the crash had not happened, no engineering argument could have convinced Air France and British Airways to ground and modify the planes. Yet, the risk was no different after the crash than it was before the crash. Was redesign of the tanks a prudent move, or an overreaction caused by the availability of the crash in memory?

Biases Due to Inability to Imagine a Result

Later in the course we'll review a sequence of events which led to the collapse of a dam on the Snake River. Investigation of the accident leads one to the conclusion that the engineers designing the dam did not believe that the collapse of a major dam was possible. As a result they made a series of bad judgments which culminated in failure of the dam.

In the oil industry, no doubt, most people would have been unable to imagine a single accident killing over 100 people prior to the Piper Alpha disaster.

Before 9/11 most people could not have imagined terrorists hijacking a plane and flying it into a building, hence clues that it was being planned were easy to miss.

If you cannot imagine an event happening, you will not recognize the warning signs that it is coming.

Illusory Correlation

In *Contact,* Carl Sagan noted that man is excellent at finding patterns in nature, even where none exist. When we see two rare events occur simultaneously we tend to believe they are correlated, even when they have no logical connection.

Years ago I investigated problems in a plant wide tank vacuum system used for emissions control. On occasion Unit A lost vacuum. Unit A operators eventually noticed that, whenever they lost vacuum, the Distribution Dept. was air rolling tanks (mixing tanks by injecting air through spargers).

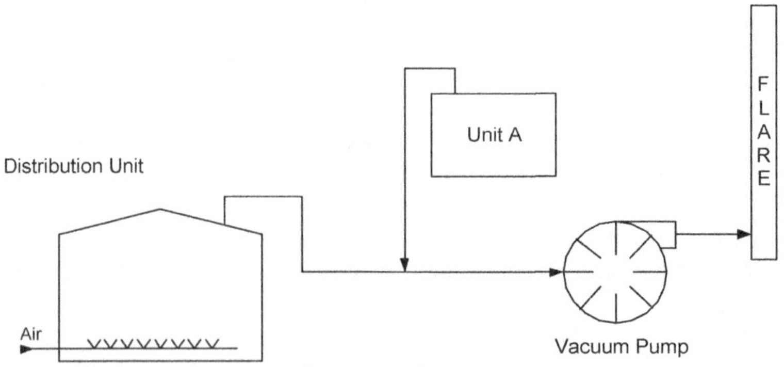

This was understandable since Distribution air rolled tanks most of the time. Never-the-less, Unit A operators became convinced that tank rolling operations in Distribution caused them to lose vacuum. Pressure drop calculations showed that this impact was unlikely. We followed up with experiments. Rolling all the tanks in the Distribution Unit simultaneously at maximum air rates had no significant impact on Unit A vacuum. Yet, the Unit A staff continued to assert that the air rolling was causing their problems until a colleague found and corrected the real problem (which was within Unit A).

The observation that they lost vacuum while Distribution was rolling tanks, caused them to correlate the two events and to believe it so fervently that the belief survived all proof to the contrary until the true cause was corrected. One result of the belief was that Unit A staff never investigated within their unit for the cause.

2.6 REFERENCE EFFECT

Is the glass half empty or half full? (Or is the glass twice as large as necessary?)

Who wants to play this game?:		
On the flip of a coin:	**Heads -**	**I'll pay you $10**
	Tails -	**You pay me $10**

Suppose I give you $10 first. Now who wants to play?:		
On the flip of a coin:	**Heads -**	**I'll pay you another $10**
	Tails -	**You give me back the $10**

Most people will not play the first game, but many people will play the second.

Difference between Games Above

The games above are identical except for the frame of reference. In the first game you will base your decision between end points $-10 and $+10 (reference point = $0). In the second game you may use the free $10 bill to adjust your reference point to $+10 making the end points $0 and $+20.

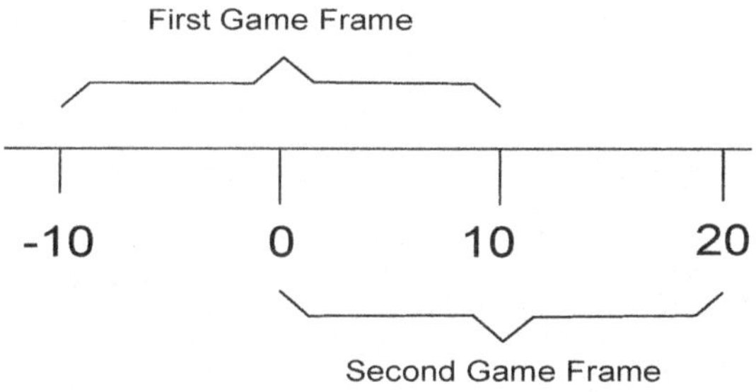

The games above suggest that we view the potential loss of money we possess more negatively than we view the potential gain of an equal amount positively. In decision problems people tend to <u>undervalue gains</u> and <u>overvalue losses</u>. This creates situations where apparently irrational choices are made. We will discuss this aspect in more detail in Chapter 4.

For now we are interested in how frames of reference might impact our assessment of data. We expect people to view data differently based on whether they interpret it as a gain or loss. But whether a consequence is viewed as a gain or a loss can depend on the way it is framed. Attempts are made routinely to influence us by framing alternatives. Auto dealers advertise $2000 off list. Jewelers frequently advertise 50% off.

Frames of Reference and Playing the Stock Market

In the stock market many people tend to "sell a winner" and "ride a loser". That behavior can be understood in the context that a loss is more painful than a similar gain is pleasurable. Hence, we hang onto a loser in the hope that it will return to the reference point (what we paid for it). A loser that returns to the black is more pleasurable than a winner making a similar advance.

Risk Tolerance for Loss

One last game; Suppose you owe me $10?		**Now who wants to play?**	
On the flip of a coin:	**Heads**	**-**	**We're even**
	Tails	**-**	**You owe me $20**

Are you more or less likely to play this game than the two games on the previous page?

Again, many people who would not play the first game above will play this game.

Does this say something about the reference effect or is this a different phenomenon?

2.7 EXERCISE: SETTING NEGOTIATING LIMITS

You are the salesman for an auto parts maker. A Japanese automaker has indicated the need for a large one time order of a part you make. This is a one-time deal with little potential for repeat business. You currently sell the part for $20 each. This is a confidential figure which the Japanese don't know. You have enough idle capacity to fill the Japanese company's order. You have determined that your marginal cost to produce the part is $10 each (you will break even on the order if you sell the parts to the Japanese for $10 each). You believe that the Japanese firm has few options and is motivated to make a deal. Discussions with the Japanese have been very difficult so far because their negotiator speaks English very poorly. At one point you thought she expressed a willingness to pay $32 each for the part. But when you quizzed her, she denied naming a price and said she wants you to propose a price at face-to-face negotiations tomorrow. Accepting that the $32 price was a misunderstanding, you agreed to the request and are now preparing for the negotiations.

Prepare for those negotiations by setting goals and limits for the following:

Your Initial Offer: $ ______/ unit

Your Aspiration Price or Goal: $ ______/ unit

Your Reservation Price: $ ______/ unit

:

Initial Offer is the price you will initially propose.
Aspiration Price is a figure you will be very happy to achieve; the lowest sales price at which you will consider the negotiations to have been very successful.
Reservation Price is the absolute lowest price you will accept. Any lower and you will walk away from the deal.

2.8 EXERCISE RESULTS

In this exercise, we have repeated part of a study performed by Whyte and Sebenius (P4). This research studied the impact of multiple anchors on group judgment. Three anchors are provided:

- the current selling price ($20),
- the breakeven point ($10)
- and the supposed offer by the Japanese negotiator.

In the Whyte and Sebenius study, different groups received different information on the price supposedly named by the Japanese negotiator. Some groups were given a value of $32 and others were given $12.

The problem contains two anchors relevant to the discussion (10 and 20) and one of dubious relevance (12 or 32). The study was designed to show that, when people are provided with an anchor, even one with little connection to the problem, they tend to use it.

Average results of the Whyte and Sebenius Study are shown below.

	Offer	Aspiration	Minimum
Low Anchor ($12)	19.3	14.7	12.1
High Anchor ($32)	32.1	22.0	17.4

Note that even the bottom line is significantly impacted. People given the high anchor claim to be willing to walk away if they cannot get at least $17.4, a full $7.4 above the breakeven point.

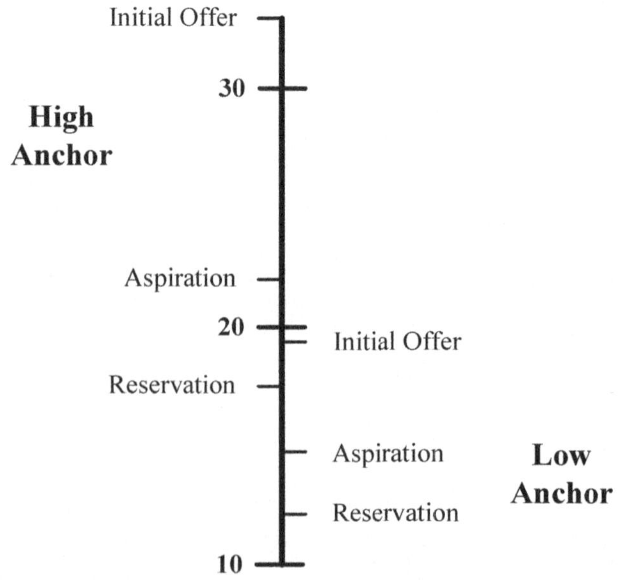

The graphic below shows a possible situation graphing the Initial Offer, Aspiration and Reservation prices for a buyer and a seller. In this case there is an overlap between the party's negotiating positions; a zone of possible agreement. Invalid anchoring can cause negotiations to fail if the result of the anchoring is that no zone of possible agreement exists.

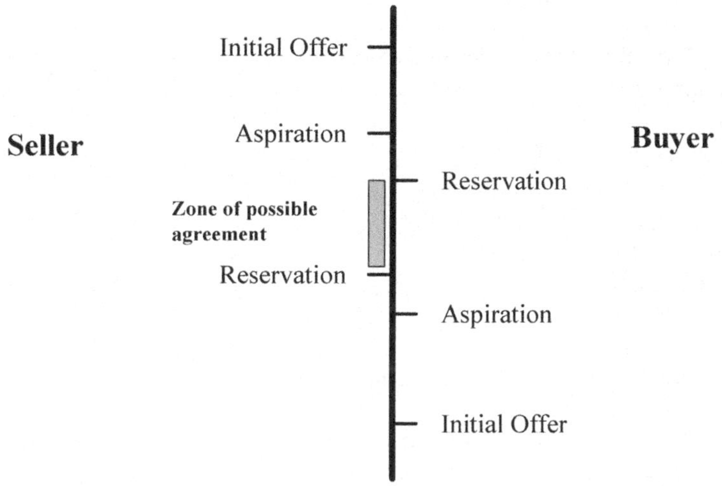

2.9 ANCHORING

When we make judgments in the face of uncertainty, we often start at some point and then adjust from that point as additional data is available. The starting point is called an anchor.

Two forms of bias are common in this process,

1) using a poor starting point, and
2) moving an insufficient amount from the starting point.

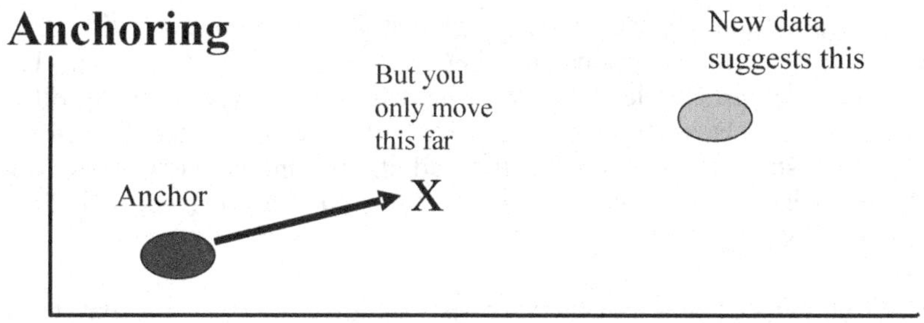

Bias of both types has been demonstrated. People tend to use any anchor available, even when the anchor has no logical connection to the problem. When additional data suggests the need for change, movement from the anchor is usually less than the data suggests is appropriate.

Anchoring Impacts all phases of decision making. We can become anchored on:

- The data we select,
- The way we interpret the data,
- The identification of objectives,
- The identification of alternatives,
- Selection among alternatives.

And there are multiple possible sources of anchors, such as:

- First Impressions.
- Intuition.
- Opinion of Authority.
- Status Quo.
- Study.
- Serendipity.
- Personal Experience.
- Opinion(s) of others.

Anchoring in Selling the Watergate Break-in

G. Gordon Liddy was in charge of intelligence gathering activities for the Committee to Re-elect the President (CREEP) in Richard Nixon's 1972 campaign. Liddy proposed a burglary of the Democratic National Committee (DNC) office. The break-in plan was risky and very expensive. When first proposed, Liddy asked for $1,000,000 which included funds for bugging the DNC, outfitting a spy plane, and parties on yachts with high priced call girls. The plan was rejected. A later plan pared down to $500,000 was also rejected. The final plan was pared down to $250,000. It was still a terrible idea, but it was accepted because it was viewed as a "good deal".

Would Liddy have garnered the $250,000 budget had he not previously established an anchor at $1,000,000?

2.10 RISK PERCEPTION AND THE AFFECT HEURISTIC

Risk Perception

One of the most important aspects of belief generation for engineers is risk perception. Risk perception differs among groups and changes within groups over time. Technical professionals seem to perceive less risk from industrial activities, including oil and gas development, than the general public does. That may be partly due to differences in knowledge. But studies have shown that additional information alone does not significantly shift risk perceptions. There are other important drivers including social forces and control issues.

And increased knowledge may not improve risk analysis accuracy. It may be that the high knowledge level among engineers makes us overly-confident in our designs and hence, to systematically underestimate risk.

Risk perception within the industry has changed dramatically over the past 30 years. The change in risk perception within the industry is a social phenomenon. Society in general has become more risk averse and workers in the industry have followed the trend.

Science and Risk Perception

Risk on projects is described as the product of consequence and probability. That is a technical construction and is not the only way that risk is perceived. Both science and probability are recent human developments. Humans perceived risk prior to the development of the scientific view. Risks are felt emotionally. And risk are defined and accepted socially.

Risk Perceptions and Heuristics

Since risk perception is a belief about the situation we should expect heuristics, as discussed in the previous sections, to play a role. For example:

- If you've recently experienced a significant accident, you will likely perceive a higher than normal risk of a similar accident occurring again (availability heuristic).
- If you have previously estimated a particular risk to be low you may continue to perceive low risk despite some disconfirming evidence (anchoring).
- People tend to be risk averse when seeking gains and risk averse when trying to avoid loss (reference effect).

Risk and the Affect Heuristic

We can add the affect heuristic to that list. The affect heuristic says simply:

> If you have positive feelings towards an activity you will tend to judge the activity to be relatively low in risk and relatively high in value. Conversely, if you have negative feelings towards an activity, then you will tend to judge the activity to be relatively high in risk and relatively low in value.

There is a natural positive correlation between our perceptions of risk and benefit. The high benefit activities we indulge in may be high or low in risk. But the low benefit activities we undertake must be low in perceived risk; else we would refrain from those activities.

But the affect heuristic points to the opposite perception – that we tend to view risks and benefits as inversely correlated; the higher the benefit the lower the risk and vice versa. The affect heuristic argues that we are likely to perceive low risk and high benefit for technologies that we think highly of and to perceive high risk and low benefit for technologies that we think poorly of.

Most people will accept orders of magnitude more risk if the risk is voluntarily accepted. The affect heuristic explains this behavior. If we enjoy an activity we will typically associate it with high value and low risk.

2.11 COPING WITH EMOTIONS

Emotions impact the way we evaluate data. If an event is very difficult to cope with emotionally, we may alter our perception of the facts. This can take two forms:

Distancing:	we deny that the event is important to us
Denial:	we deny that the event occurred

Distancing

Distancing is very common behavior. In the early stages of personal relationships both parties jockey for position in order to avoid being hurt. If things turn out badly, they can argue that they really didn't care anyway.

Joking is sometimes a form of distancing. When something is troubling, you don't have to take it so seriously if you can joke about it and others laugh.

Distancing is not always a problem. Medical doctors use distancing as an occupational coping mechanism. It would be difficult for them to practice if they became emotionally involved in the pain of every patient.

Denial

Denial is different from distancing. In distancing we accept the fact that the event occurred, but minimize its meaning. In denial, we deny the fact.

These insights about emotions are taken largely from the work of Lazarus and Lazarus (12).

2.12 SOCIAL PRESSURE

Social forces impact the way we evaluate data. The cases and arguments below are taken from Cialdini (13).

Doomsday Cult

1964, Chicago Illinois. A middle aged man and his wife led a cult with about 30 members. The wife claimed to have the gift of channeling with the spirits; they told her things. She wrote them down and shared them with the group. The voices told her that, because it was full of evil, the world would be destroyed by a flood (not a very original story). She has also been told that they, and they alone, as the true believers would be saved by a flying saucer (high-tech ark). The flying saucer was to arrive at midnight on a specified day, and about daybreak of the following day the catastrophic rains would begin.

The group members made no secret of their beliefs and their actions were consistent with their espoused beliefs; many gave away all their property. But they did not attempt to convert others and they didn't seek publicity. Newspaper and TV reporters repeatedly tried to extract interviews and were repeatedly turned down. The group was too busy making preparations.

On the evening of the scheduled arrival of the flying saucer they gathered together to pray and wait. Midnight came and went. Initially there was quiet; then quiet fear; crying; wondering. Around 4:00 the wife received a message; "Because they have believed, the world has been spared".

Then a strange thing happened. The cult, which until then had shunned publicity and which had not sought to make any converts, suddenly started calling all the local newspapers, TV stations and radio stations trying to tell their story. It was a good story about how they saved the world. Further, they began actively trying to convert their friends and neighbors.

Why? Why seek publicity at the very moment when their predictions have failed? Why seek converts at the moment when their credibility was weakest?

Arguments: Social Proof

We seek proof for the things we believe. Scientific or rational proof is preferred by most people. But social proof is also very powerful. If many people believe something, we are more likely to believe it. If we can convince others to believe what we believe, we feel better about believing it and more justified in believing it.

In the case of the doomsday cult, as long as they could believe the flying saucer was coming they had no need for social proof. But then their facts were proved wrong; no flying saucer came. They could not simply stop believing, they had invested too much into this "truth". Denied of rational, scientific proof, they sought social proof instead. If they could convince others, they could feel justified in continued belief.

Social Proof at Work

Social proof is prevalent at work as well. People on a given team tend to believe many of the same things. When team members switch to other teams they will, over time, be converted over to that teams points of view.

I worked in a benzene separation unit early in my career at a time when the allowable exposure limits to benzene were dramatically lowered. The Industrial Hygiene (IH) Department launched data collection efforts in the unit and imposed operating rules that seemed onerous and unnecessary to those of us responsible for running the unit. When a good friend of mine was promoted to IH Department Head I expected a change in direction. But my friend quickly adopted the views of the IH Department.

2.13 ON SELECTING AND EVALUATING DATA

Data Collection is Typically Haphazard

A large part of the effort in design of scientific experiments is in determining the data to be measured. A well designed data collection program would include consideration of at least the number of independent measurements, and the required accuracy of the measurements to establish confidence in the results. But in engineering design work, well designed experiments are the exception. Usually, we collect data much more haphazardly with little thought for variability.

Often we only collect the data that attracts our attention. What attracts our attention depends on factors such as salience, vividness and relevance to our models or expectations.

Impact of Heuristics on Data Collection and Evaluation

The heuristics we've been discussing affect which data we choose to collect, which we choose to ignore, and how we evaluate collected data:

- We may not measure an important variable thinking that it is not important or that we already know the value of it (availability).
- We may avoid collecting data that could be threatening or embarrassing (distancing).
- We tend to make too much of too few data points (illusion of validity error).
- Once having assumed a position, new data is not adequately integrated into our beliefs (anchoring, sensemaking).
- We may conveniently ignore or diminish data that does not fit our beliefs (distancing, denial).

Model of Expectations

So far we have discussed how preconceived notions impact data collection and evaluation. The opposite situation, collecting data prior to knowing what the data might reveal, is also problematic.

I remember wondering in college why we spent so much time modeling experiments and predicting the results before doing the experiment. It seemed backwards. Why not run the experiments first, collect the data, and see what it said.

Without some model of expectations it is difficult to judge the meaning, or the validity of the data collected. Problems with the experimental procedure are often detected as violations of expectations. If the answer you get is different than what you expected you will likely search for an error or redo the experiment.

Many of the errors made in computer calculations can be avoided by anticipating the result prior to running the program.

Don't confuse me with the facts.

The data you collect is not the real data. It is a sample of the real data. At best it is a random sample. Usually it is not. Often it is a highly biased sample "designed" to justify what you already "know" to be true.

2.14 ASSIGNING MEANING – CREATING INFORMATION

I've played fast and loose with the word "data" so far. "Data" and "information" are typically used synonymously. But there is an important distinction.

Data

Data is observable, measurable, recordable phenomena. Data is cold, hard, raw. Examples of data are voice which can be recorded and stored via tape recorder, instrument readings which can be recorded in computer files or manually recorded on paper by an operator, etc.

Assigning meaning to data

Data has no meaning. Everyone will agree on the data, but individuals may disagree on what it means. We assign meaning to data. Data is information if it can be assessed against a set of possible alternatives. A pressure transmitter reading indicating a vessel operating pressure of 50 psig is data. If you know that the normal operating pressure is 35 to 40 psig then you can assign meaning to the data; the vessel is operating at an abnormally high pressure. That is information.

Example: "Jane is a woman"

This simple statement will be interpreted in most situations to mean that Jane is female human of adult age. We assume a conventional, literal meaning, barring some context in which another meaning might be deemed appropriate. The sentence can have other meanings depending on context. For example:

> If we know that Jane is a child, then "Jane is a woman" has a second order meaning. She is perhaps very mature for her age, or looks older than she really is.
>
> If used in the context when other women her age are called "ladies", then the phrase might mean that Jane is crude and unsophisticated.

The sentence "Jane is a woman" is information only if you can assign meaning to it.

2.15 SENSEMAKING

How can I know what I mean until I see what I say?

We are capable of believing things with a minimum of supporting data or even no data if the belief is salient or if there is a social advantage in believing. We often develop arguments to justify positions that we have already taken. The development of arguments to justify existing beliefs is a form of sensemaking.

This classic account of sensemaking captures the essence:

> "The little girl had the making of a poet in her who, being told to make sure of her meaning before she spoke, said "How can I know what I mean until I see what I say?".
>
> from *Sensemaking in Organizations*, Weick (5)

Making Sense of Christopher Columbus

Christopher Columbus was a hero in his time and for most of the next 500 years. In the past few years his treatment of native populations in the lands he "discovered" has been heavily criticized. A new generation of people with different experiences and values is making sense of Christopher Columbus' exploits in a different way.

The story hasn't changed. Current events have changed the meaning of the historical data. A new model has been developed for judging explorers. Columbus has been judged against this new model and found wanting.

Sweater Knitters Making Sense

In the area of Hawick, Scotland, there are 17 firms manufacturing high quality cashmere sweaters. Because Hawick (pronounced oyk) has a large pool of skilled workers, they continue to produce sweaters by hand. Their methods are not as efficient as more modern techniques (electronic knitting machines) adopted by competitors in other parts of the world. Hawick sweaters are more expensive. They sell them in specialty shops and department stores dealing in high quality sweaters.

People in Hawick have been making sweaters this way all along. But they didn't always make expensive, high quality sweaters. Their sweaters became high quality and expensive when other manufacturers started making cheaper sweaters via machines. What has changed over time is the sense they make of their business. They did not originally begin making sweaters by hand in order to produce higher quality sweaters for which they could charge higher prices. They did not interpret their actions as the pursuit of high quality until price became an object.

The people of Hawick take credit for their foresight in creating high quality handmade sweaters, yet it was not foresight, but hindsight that generated the current situation.

Sensemaking in Athletes

Another powerful instance of sensemaking occurs on sports fields. Star athletes are usually self-confident. This self-confidence is based on the affirmative reactions of others to their exploits. They feel good about themselves because others feel good about them.

2.16 PROPERTIES OF SENSEMAKING

Sensemaking is Retrospective

Sensemaking is a retrospective reconstruction of history. Retrospection simplifies the history of complex activities. If people know the outcome of a complex prior history, they may make sense of that outcome as the inevitable result of flawed analyses and incorrect actions, even when those actions were not influential in the outcome and the likely impacts of those actions were not at all obvious a priori. Retrospection often wrongly implies that errors should have been anticipated and avoided.

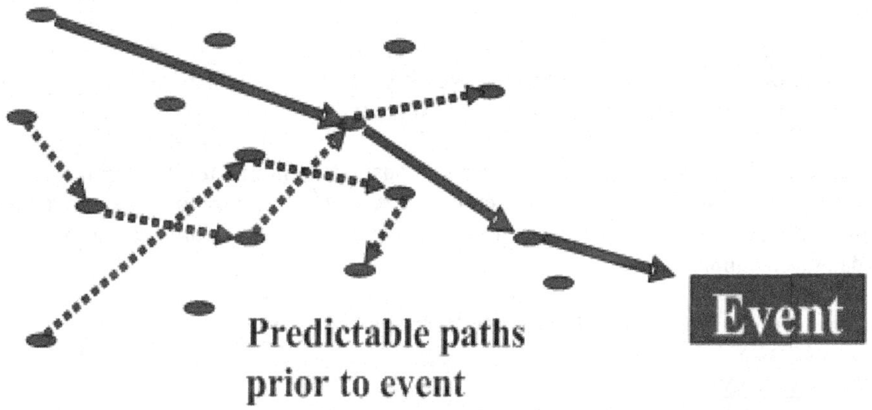

A remarkable and well researched incidence of this is the Space Shuttle Challenger disaster. The Presidential Commission that studied the event provided a simple history of the event identifying the causes and blaming the NASA organization for failed processes and analyses. In *The Challenger Launch Decision,* (15) Diane Vaughan relocates the key players in the working environment before the event and shows that the flaws "discovered" by the Presidential Commission were not at all obvious before the fact. The investigators found what they expected to find.

Sensemaking is Social

The Hawick sweater makers could not believe that they make high quality sweaters if customers weren't buying them. The Hawick sweater maker's concept of themselves is grounded in the perception that others have of them. They cannot believe that they make high quality sweaters unless the public believes that they make high quality sweaters.

"How can I know what I've made until I see what others say about it?"

Athletes can not feel special unless the crowd cheers.

"How can I know who I am until I see how they respond?"

Sensemaking is both Product of the Environment and Creator of the Environment

We tend to view our environment as fixed. But the environment does not simply exist as a given to be dealt with. The environment is a product of our sensemaking and we are a product of our environment. It is an iterative process. We influence our environment as it influences us.

Early in my career I noticed that projects on which I had a good operations representative went smoothly and projects on which I didn't have an effective operations rep went poorly. I felt that I was at the mercy of chance; either I got a good rep and the project went well or I got a poor one and the project went poorly. It was some years before I realized that I could have a large impact on the effectiveness of the operations reps assigned to my projects. By working closely with the operating reps, working hard to understand their objectives and making clear what I needed from them I could significantly impact their performance. The environment was not a given.

Sensemaking is Usually Limited by a Lack of Mental Models Rather than of Data

We tend to believe that we can't make sense of an event because we don't have enough data. But the problem with making sense of past events often is not missing data, but lack of values and models with which to evaluate the data.

Estimation of the pressure drop in piping was problematic until Fanning and Darcy developed the concept of friction factor and related it to Reynolds Number and pipe roughness. Until they developed a representative model of pressure drop in pipe much of the data didn't make sense.

Sensemaking is Driven by Plausibility Rather than Accuracy

Will any map do?

> *A small Hungarian military detachment was on maneuvers in the Alps when a blizzard hit. It snowed for two days. They were lost, cold, hungry. On the third day the weather broke, but they were lost. Then one of them found a map in his bag. With the map they found their bearings and walked out of the Alps to safety. They told their story to their lieutenant upon return to base. The lieutenant looked at the map and discovered that it was a map of the Pyrenees, not the Alps.*

Could it be that any old map will do when you are lost? The value of this map was that it calmed the soldiers down and motivated them into action. Their action gave them a chance. Action generated other cues for them to respond to. It is not plans, but actions that generate results. Actions are motivated by the plausibility of success.

We need to know enough to get on with our projects, but no more. Sufficiency and plausibility take precedence over accuracy and completeness. From the standpoint of sensemaking with data, it is usually more productive to filter available data and make sense of the world using some simplified model, than to painstaking seek definitive accuracy.

There are many reasons why this must be so. For starters, people need to filter data in order not to be overwhelmed by it. To be effective one must sacrifice some accuracy for speed. A sufficient and plausible answer provides the necessary impetus to spur action.

A second reason why sensemaking must be focused on plausibility is that collecting more data does not always make one more certain of the right course of action. Often data can have multiple meanings. The meaning we land on is often governed more by our mental models than by the data. It may not be possible, at the time of perception, to tell whether the perception is accurate or not. Many errors only become so in hindsight.

Accuracy is often secondary to speed because business information is time sensitive. Speed also often reduces the need for accuracy. A quick response shapes the environment which then impacts what the next reasonable action might be. Waiting until you are sure is a hopeless game of catch up.

Accuracy is project specific. Actions that result in successful outcomes on one project will not necessarily be successful on another project. We can never say that a given action, in a non-trivial setting, will result in a given outcome with 100% certainty. Success frequently depends on many actions by many people coming together in effective ways in a suitable environment, and often depends on some amount of luck.

A final reason why plausibility is more important than accuracy in sensemaking is that the stimuli that are filtered out are often those that inhibit bold action. An accurate assessment of a difficult situation will tend to dissuade sensible people from taking action. A plausible, though optimistic, assessment may motivate action.

A Positive View of Heuristics

The early sections of this chapter painted a largely negative view of heuristics and biases. The notion that plausibility is more important than accuracy conflicts with that view. Heuristics such as representativeness, availability, and anchoring were presented as sources of error. Indeed they can be.

But they are also the tools that allow us to rapidly make sense of complex situations and to act reasonably with a minimum amount of data. As such they are also very significant strengths.

Information in this section is largely from Weick (5).

2.17 DISSONANCE REDUCTION

Making Sense of a Difficult Decision

Sensemaking occurs in the aftermath of any difficult decision. As you attempt to decide between two equally attractive alternatives you may flip-flop on your preference for a period of time. You can name pros and cons for each alternative.

When you finally do make a decision, the act of making the choice will create some anxiety (dissonance); you will be giving up some attractive attributes of the rejected alternative and accepting some undesirable features of the selected alternative.

After making the decision you will begin making sense of the decision in order to reduce your anxiety. You work to make the selected alternative look better and the rejected alternatives look worse. Dissonance reduction consists of spreading the options and can take several forms:

Accentuating Selected Objectives:

> One way that we spread the options is by putting emphasis on the positive aspects of the chosen alternative and/or by putting emphasis on the negative aspects of the rejected alternative.

Rationalization:

> After having made a decision you may come up with reasons for your choice that did not actually play a role in your selection. This is especially true if you can identify altruistic reasons for the selection you made.

Commitment:

> Making a firm commitment to a course of action minimizes regret over the course not chosen.

Defensiveness:

> To mask their anxiety people may become defensive of their choice.

Note the existence of a reference effect here. It is much harder to influence someone after they have made a decision and especially after they have publicly stated their decision, because they now view the selected option much more positively relative to the rejected option(s).

2.18 SENSEMAKING IN ORGANIZATIONS

It is reasonable to say that Isaac Newton created the science of physics by identifying and achieving acceptance of the core principles of the science. In *The Structure of Scientific Revolutions*, (37) Thomas Kuhn argues that a field becomes a science when the body of theory for the field becomes widely accepted. A key feature of this acceptance of the fundamental theories is that it then restricts future research. Research conducted within the science is restricted to areas that make sense within the accepted body of beliefs. No physicist can get funding to study the impact of anger on the force of gravity, for instance. It is firmly believed by all physicists that there is no such effect and that no effort should be devoted to such study.

Sciences Change Pragmatically

But the rules of a science do change over time. The rules of physics have changed, or been expanded, significantly over the past century. How does that happen? Sciences change via a continuing attempt to decrease discrepancy between accepted theory and observations. When a well-accepted theory no longer adequately explains experience, doubt is created. Efforts are taken to dispel the doubt. When a better explanation/theory is found the scientific community eventually accepts the change (though sometimes only as the adherents to the old theory grow old and die). Sciences change as scientists make sense of the new experiences.

Organizations Change Pragmatically

Within the field of communications theory there is a field of study called Pragmatics (38). Pragmatics is the study of way in which context contributes to meaning. Pragmatists study how communication creates organizations and how those organizations progress. Their work shows that organizations change the same way that sciences change, they change pragmatically; they change via a continuing attempt to decrease discrepancy between accepted theory and observations.

Organizations are complex entities often consisting of many people, formed into multiple groups with hierarchies and more or less agreed roles and responsibilities. Some obvious questions arise when we look at a complex organization:

- How did it get established?
- How were roles and responsibilities assigned?
- How did the people in positions of authority achieve that authority?
- Why do the other people in the organization accept the established authority and roles and responsibilities?
- What does it mean for an organization to have objectives? How does the organization motivate individuals to work towards those objectives?
- How does the organization learn and change?

Taylor and Van Every (38) discuss all these issues and conclude that communication plays the pivotal role.

Organizations are in a constant state of flux. People in the organization continually make sense of new experience and communicate with others to have their interpretation of experiences shared by the overall organization. It is a game and the fundamental objective of the game is control. Individuals achieve authority by succeeding in having their perspective accepted as <u>the</u> perspective of the organization. The person(s) who thusly achieve a measure of authority then author the organizations objectives, strategies, processes, etc.

One would think that the losers in this game might refuse to play anymore. Why do people in organizations accept subordinate roles? It must be because they get something out of it too. Organizations provide income, stability, prestige, authority over some others, the potential for wielding authority in the future, potential to accomplish things greater than one could achieve alone.

A successful organization is one that has achieved general agreement on objectives and roles and responsibilities including authority and in which the individuals at all levels are sufficiently motivated to work towards the objectives of the organization.

How does an organization take action to achieve its objectives?

We engineers tend to think that we build bridges and buildings and oil production facilities. But few of us wield wrenches, crimp wires or weld plates of steel together, etc. An engineering organization takes action to achieve its objectives by creating documents that instruct others (vendors, welders, mechanics, etc.) to take appropriate actions.

Pragmatists focus on the documents that an organization generates. Documents are particularly important in engineering organizations, because they are our product.

How do we Generate Design Documents?

We generate them pragmatically, of course. They are written initially by subordinates. They start out rough, full of errors, then are reviewed and improved by more senior staff.

I think of it as a document distillation column. The document is refined as it moves up and down the review column. At each stage in the column technical errors are identified and removed. But the review has another function – to strip out heresy. There is also a control game being played. Every grunt wants a better life. They jockey to have their opinions accepted. Those who win this control game will rise up the organization in future projects.

2.19 SENSEMAKING IN CHALLENGER LAUNCH DECISION

Sensemaking in Accident Investigations

Suppose you consider an organization to be very well run, but then a serious accident happens. There is one thing you can bet on: the accident investigation will find management incompetence or wrong-doing.

When an accident happens in a well-run firm we are forced to make sense of it. The accident creates a conflict: well-run firms aren't supposed to have accidents. The simple first step in making sense of such a situation is to hypothesize that the company may not be so well run after all. Investigation will find OSHA violations (many firms that are inspected have violations). The investigation will find emails pointing out a defect in the machine that broke (ignoring similar emails about machines that didn't break and contradictory emails from others who believed there was no problem). Operators will be shown to have routinely violated operating procedures (operators do that everywhere).

In short, no matter how impressive a company's management is, it will likely be possible to craft a case of management neglect or incompetence. It will almost certainly be possible to craft a story of the sort "If they had only done ____, then the accident would not have happened." A good example is the Challenger accident investigations.

The Challenger Launch Decision

The Shuttle Challenger "exploded" on takeoff due to a joint leak in one of the Solid Rocket Boosters (SRB). The SRBs contained joints which were sealed by two O-rings as shown below.

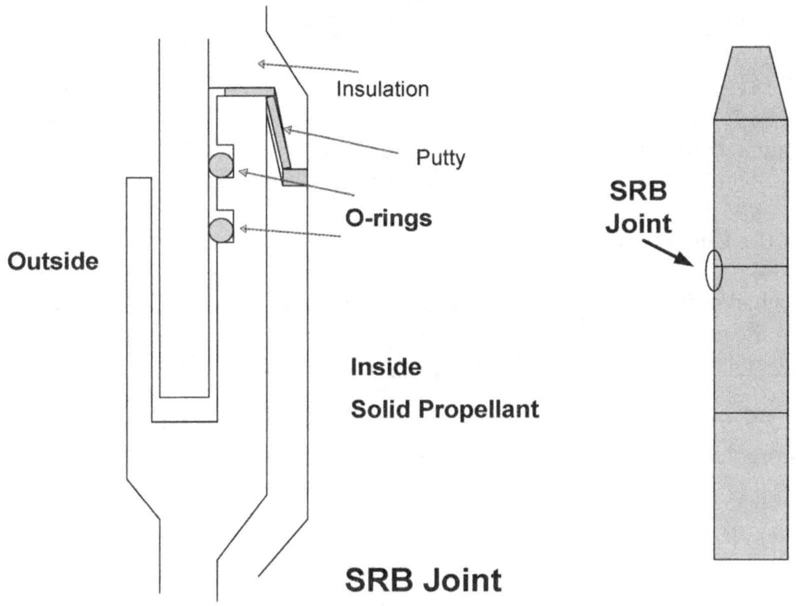

Prior to the Challenger launch several flights suffered damage to the primary O-ring and one suffered charring of the secondary O-ring. The damage appeared to be more severe at lower temperatures. The temperature on launch day was in the low 20's °F, the coldest launch day on record by far. The coldest previous launch occurred at 53 °F.

Investigations after the crash concluded the following:

Presidential Commission:

- NASA production pressures caused them to ignore safety issues in order to get in more launches.
- Flawed decision making processes.
- Inadequate procedures for reporting safety issues.
- Violation of reporting rules by the Thiokol (SRB manufacturer) and Marshall Space Center managers.

US House of Representatives Review

- People had been making poor technical decisions about the O-rings over a period of several years and they "failed to act decisively to solve the increasingly serious anomalies in the SRB joints."

Speculations in the Media:

- President Reagan's State of the Union address to Congress was scheduled for that night. The media speculated that there was pressure on NASA to launch prior to the speech in order to get favorable mention in address.
- NASA got bad publicity over frequent delays of the Columbia shuttle launch a few weeks earlier. Media speculated that they were eager to avoid a repeat.

Vaughan's Research

Diane Vaughan made an in-depth investigation into the events surrounding the Challenger launch decision (15). The following discussion is taken from her work.

Viewpoint of Investigators

The investigators concluded that the accident resulted from management wrongdoing because the suspect SRB was flown repeatedly for years despite documented problems.

Viewpoint of the Engineers

The engineers had a different view. Several times the behavior of the joint surprised the engineers. At those points they expressed concern about the joint. <u>But in every case up to that point, further study convinced them that the joint was an acceptable safety risk</u>.

The joint never performed as the design engineers intended. Over the course of several years several problems were discovered.

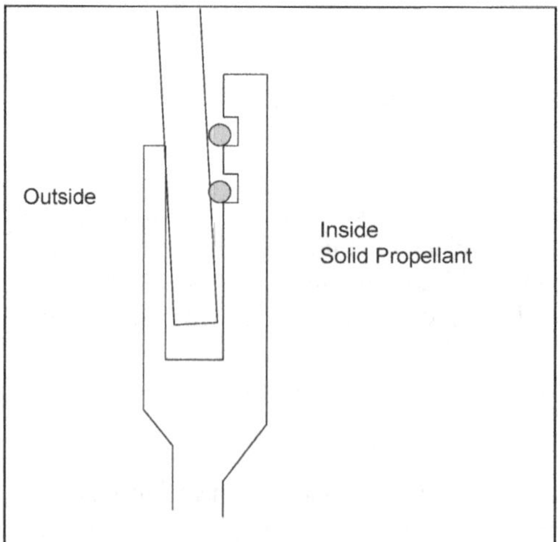

Joint Rotation Caused by Internal Pressure

dynamic effect, short-lived (0.6 seconds)

The most significant was joint rotation caused by internal pressure. This rotation caused the O-rings to come off their seats. The phenomenon was discovered before the first flight.

Detailed evaluation showed that the phenomenon was very short-lived (fraction of a second) and the engineers concluded that it was an acceptable risk.

Problems were noted (charred or burned O-rings) after several launches. In each case investigation led engineers to accept the joint design as safe to fly.

This pattern, repeated several times, became part of the group's culture:

- Signals of potential danger.
- Official acknowledgement of potential problem.
- Review of evidence, analysis, testing.
- Conclusion that joint performance was acceptable.
- Acceptance of risk – Shuttle launch.

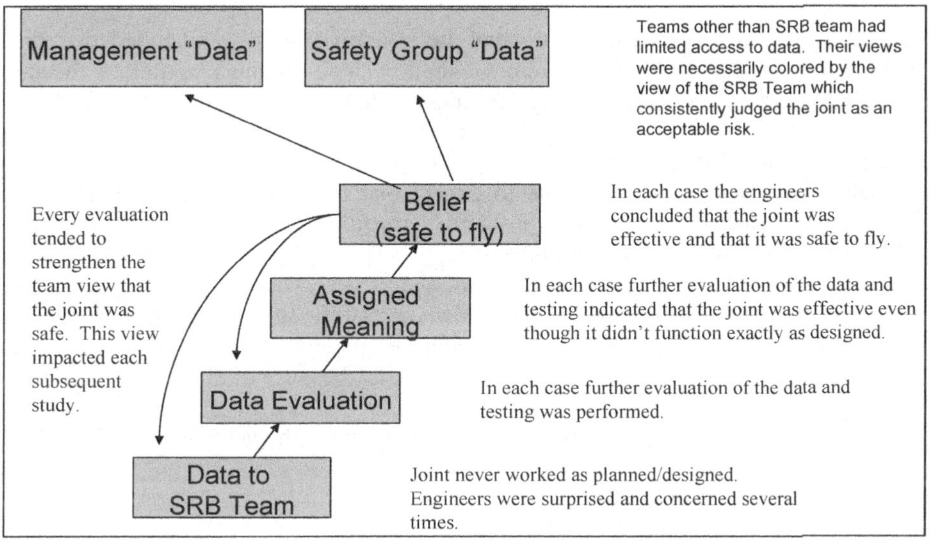

Chronology - Weeks before the Flight

In the weeks leading up to every flight NASA conducts very thorough safety reviews. NASA was criticized for not covering the SRB joint issue in these meetings prior to the Challenger launch. But the meetings are designed specifically to address new information. There was no new information on the joint to discuss. The criticism was unfounded.

The Day before the Flight

In the afternoon of the day before launch, Thiokol engineers expressed concerned about the low temperatures, much colder than any previous launch. NASA asked for a recommendation on a minimum launch temperature. Thiokol engineers suggested not flying below 53 °F (the lowest previous launch temperature). NASA, understandably, pushed back "you want me to wait til spring to launch!?".

Thiokol could not defend a temperature limit of 53 °F and they could not craft a story resulting in any other selection of a minimum launch temperature. (How cold is too cold?) There was criticism in the press and by investigators that NASA pushed Thiokol to prove that it was unsafe to launch, the opposite of previous behavior in which they insisted that Thiokol prove that it was safe to launch.

In fact NASA did insist that Thiokol prove flight worthiness. But that had already happened in the weeks prior to launch. What happened the day before the launch was different. Here we had a group of concerned engineers trying to alert management to a somewhat ambiguous concern, not fully thought through, and having to fight through the organizational structure to be heard. Probably cold temperature concerns had not been fully considered previously because it is very rarely that cold at the Cape Canaveral launch site. The Shuttle had already been proved flight-worthy. There was inertia (anchoring) supporting launch. The burden of proof was largely on the concerned engineers.

Learning the Wrong Lessons

The Presidential Commission and other investigative bodies concluded that the rules were inadequate and that rules were violated. These findings suggested the need for more bureaucracy. NASA took these findings to heart and made changes.

Vaughan concluded that the bureaucracy itself was a large part of the problem. If Vaughan is correct it is reasonable to expect that the changes implemented by NASA would be ineffective at best and probably harmful. A decade later the Kraft Report (1995) concluded exactly that, noting:

> "As a result of the Challenger incident a "safety shield" philosophy has evolved creating a difficult management situation. Managers, engineers, and business people are reluctant to make decisions that involve risk because of the fear of persecution."
>
> and "…the <safety program> used today may make the vehicle less safe…."

NASA learned the wrong lessons.

Structural Secrecy

Probably the most significant lesson to be learned from the Challenger decision is the impact of structural secrecy. Organizations are structured around a division of labor model. People specialize in certain skills. No one in the organization knows the whole story. Stakeholders and experts in this case were divided between the Thiokol Office in Utah, Marshall Space Center in Alabama, the launch site in Cape Canaveral and NASA Mission Control in Houston. Managers had only pieces of the story. Marshall Space Center managers would have cancelled the launch had they fully understood what the Thiokol engineers were trying to tell them.

2.20 GROUP IMPACT – STRUCTURAL SECRECY

Organizational Structure

A typical organization is composed of multiple teams of specialists. Each team typically interfaces, formally or informally, with other teams at the same level in the hierarchy while reporting formally up to the next level of management.

Data Collection and Transfer

Individuals collect raw data, evaluate it (add meaning) and then transmit summaries and inferences to other teams. The collection and evaluation of data is not unbiased. Team positions and priorities impact what individual team members believe, what data they collect, how they evaluate the data and what they report. Discussions within teams may be relatively open and honest, but team-to-team and team-to-management communications are likely to be highly filtered and designed more to achieve team objectives than to transmit facts and information. Several factors conspire to dilute the effectiveness of data transfer including:

Ignorance of others needs and knowledge:

If you don't have a good grasp on what members of other groups do and why they do it, you will not be sensitive to their information needs and you will not be sensitive to the impacts that your decisions have on their tasks. You may not pass along the information they need simply because of your ignorance of their needs.

Even when you know that members of other teams need some particular piece of information, you may not provide it if you think they already know it. This is particular true of tacit knowledge. It is very difficult to know what others know and what they don't, especially if the working relationship is distant. Everything you "know" you've inferred from data. Others with the same data may not "know" what you "know" and may well "know" something very different.

Control Issues

Data is power. We all seek control. We cherry pick the data that supports our positions, we evaluate the data to our group's advantage and we pass along this skewed analysis. When data embarrassing to our team is found, it is likely that that data will not be provided to other teams or at least not provided in a timely manner.

Management Dilemma – Structural Secrecy

We depend on others to provide information that we need to make decisions within our area of expertise. And we provide information to others so that they can make informed decisions in their areas of expertise. But the information shared by each team is likely to be biased towards achieving that team's objectives and/or to avoid embarrassment.

We should expect management to routinely obtain biased and limited information. More specifically we should expect managers to hear, from each team, information that is calculated to promote that team's objectives. Also, we would expect that management would rarely hear "bad" news or embarrassing information from below.

Structural Secrecy is hard to Combat

Structural secrecy is built into the very structure of organizations. To battle these problems, organizations frequently set up reporting rules. Ironically, these rules can make the situation worse as the organization gets flooded with paperwork that no one reads and which is often written to control rather than to inform anyway.

2.21 CHAPTER SUMMARY – UNDERSTANDING MISUNDERSTANDINGS

Why do you believe the things you do?

You may believe things for a number of reasons including:

- Scientific evidence.
- Salience.
- Social pressure.

It is possible to believe things with little or no direct evidence, or even in the face of contradictory evidence, if the belief is salient and/or if there is social pressure to believe.

To assess situations and develop beliefs we use heuristics. Heuristics speed data evaluation (as compared to rigorous statistical evaluation), but can cause biases. Among the more common heuristics and biases are:

- Anchoring – being stuck in place based on previous assumptions or data.
- Availability – The tendency to believe that certain events are more or less likely due to the availability of similar instances in memory or the ability to imagine them.
- Reference Effect – The tendency to believe that if A is similar to B in some ways it is similar to B in other ways as well.
- Affect Heuristic – If we think highly of something we will underestimate its risk and overestimate its benefits.

You have preconceived notions based on your previous experiences. You believe the things that you do because of your particular history and social group membership. This history impacts the selection of data you collect and affects the sense you make of data.

Others with a different history may see the same data, but believe different things. It will often not be possible to achieve agreement with people who have a different history and or different allegiances. Since your opinions seem obviously correct, the differing opinions of others will seem wrong. You will naturally make assumptions about the reasons these people hold erroneous beliefs:

- they may be biased,
- they may have nefarious objectives,
- perhaps they aren't very smart, or
- maybe they suffer from poor judgment.

These opinions of others make it even more difficult to work through differences.

2.22 ANSWERING THE FOUR QUESTIONS

At the end of chapters 2 through 8 we will try to generate answers to the four questions posed in chapter 1:

- ***Why do so many changes occur late in projects?***
- ***Why do we have so many problems at interfaces?***
- ***Why do we repeat past mistakes?***
- ***Why do we so often finish behind schedule?***

What have we learned in this chapter?

Why do so many changes occur late in projects?

Flawed beliefs caused by biases in data evaluation cause us to choose poor alternatives. These erroneous beliefs may be due to availability error – the inability to imagine something happening, or the illusion of validity – confidence based on too few data points. But errors, whatever the cause, are not the issue here. The issue is our failure to correct such errors until late in the game. Why do we hang on to flawed designs?

<u>Anchoring:</u>

We should expect to see significant anchoring in engineering designs. Once you've selected a design it is difficult to change.

<u>Sensemaking:</u>

Once we have anchored on a design, we are capable of defending it by focusing on its good features and rationalizing away its problems. Having selected a design we will spread the alternatives (dissonance reduction) so that over time the selected alternative becomes more attractive.

<u>Designs 'frozen' too early</u>

It is common practice to freeze aspects of the design at points in time on a project. This is necessary to give the various design teams a common design basis and avoid surprises. But if the design is frozen too early many errors are "frozen" and these are then difficult to correct.

<u>Frame of reference error (sunk cost):</u>

The more you invest into a selected option the harder it is to change.

<u>Risk is underestimated due to familiarity (affect heuristic)</u>

The design team will have a high opinion of the design and will naturally over estimate its positive aspects and underestimate its negative aspects including risks.

Why do we have so many problems at interfaces?

<u>Failure to share data/information:</u>

This can happen for several reasons. Structuring organizations into teams creates structural secrecy which limits the flow of data, especially data embarrassing to the team that developed/discovered it. Some data is not shared simply out of ignorance of other's needs for the data.

<u>Data means different things to different people:</u>

Much of what we call data is actually meaning applied to the data. Different people infer different meaning from the same data. This may be accentuated by team membership as people use data to support their team's positions.

<u>Social pressure to support the team:</u>

We expect our team members to support us, even when we are wrong.

Why do we repeat past mistakes?

<u>Anchoring on past success:</u>

A practice that worked in the past is difficult to abandon.

<u>Sensemaking:</u>

History is more complicated than history books suggest. It is usually possible to craft more than one story to explain a poor result. It is easy to "learn" the wrong lessons.

<u>Salience:</u>

In *Flawed Advice and the Management Trap* (1) Argyris argues that most business advice is bad advice. One of the reasons cited is that much of the published business advice seems so obvious that we don't seek any proof or justification that it will work. It is difficult to challenge practices that seem obviously correct.

<u>Structural Secrecy:</u>

Managers who could affect meaningful change are shielded from the true causes of error by the organizational structure.

Why do we so often finish behind schedule?

<u>Sensemaking of the future:</u>

In the same way that we simplify the past, we also simplify the future. Sensemaking of the future results in overly simplistic projections and optimistic schedules.

<u>Structural Secrecy:</u>

Schedules are important to project management. Schedules are often optimistic because management puts pressure on subordinates to make optimistic projections. When problems occur subordinates are often aware of schedule problems long before the project managers are.

<u>Collecting too much data:</u>

It is certainly possible to get carried away in data collection and generation. Twenty years ago we ran process simulation programs a few times on each project to generate the design basis and we shared this data widely. Now we run simulation programs many thousands of times on a single project and it is often unclear what data should be used.

<u>Data means different things to different people:</u>

When we attempt to make decisions with multiple stakeholders we have difficulty when different people interpret data in different ways.

<u>Dissonance reduction is time consuming:</u>

Making a difficult decision is time consuming, and the hard work isn't even finished with the decision. Hard decisions create anxiety. Anxiety reduction takes time and action may be impeded during this period.

Chapter 3

Setting Objectives and Identifying Alternatives

Sources:

The most important information in this chapter is derived from the work of Ralph Keeney as summarized in *Value-Focused Thinking* (2). Value-focused thinking is a structured approach to using your values to guide decision making.

Additional information including the development of objectives structures comes from Robert Clemens, *Making Hard Decisions* (8).

Preface:

We are action oriented people. When confronted with a problem we begin immediately to seek alternatives to correct the problem. Often we do that before we even fully understand what the problem is. This focus on alternatives and action often results in our solving the wrong problem for the wrong reasons.

The solution is to change your focus from alternatives (alternative-focused thinking) to objectives and values (value-focused thinking). Value-focused thinking is a structured approach to identifying your objectives and identifying alternatives that satisfy your objectives.

The number one problem with learning to play golf is our natural tendency to do the obvious thing to solve a problem.

Percy Bloomer

CHAPTER 3 CONTENTS

3.1 INTRODUCTION
3.2 EXAMPLES OF POOR OBJECTIVES IDENTIFICATION
3.3 ADVANTAGES OF VALUE-FOCUSED THINKING
3.4 OBJECTIVES STRUCTURING
3.5 TECHNIQUES FOR IDENTIFYING OBJECTIVES
3.6 USE CASES
3.7 EXERCISE – IDENTIFYING OBJECTIVES
3.8 EXERCISE – EXAMPLE RESULTS
3.9 ATTRIBUTES – MEASURES OF SUCCESS
3.10 TYPES OF ATTRIBUTES
3.11 CONSTRUCTED ATTRIBUTES
3.12 PROXY ATTRIBUTES
3.13 OBJECTIVE FUNCTION
3.14 USING OBJECTIVES TO IDENTIFY ALTERNATIVES
3.15 INFLUENCE DIAGRAMS
3.16 CHAPTER SUMMARY
3.17 ANSWERING THE FOUR QUESTIONS

3.1 INTRODUCTION

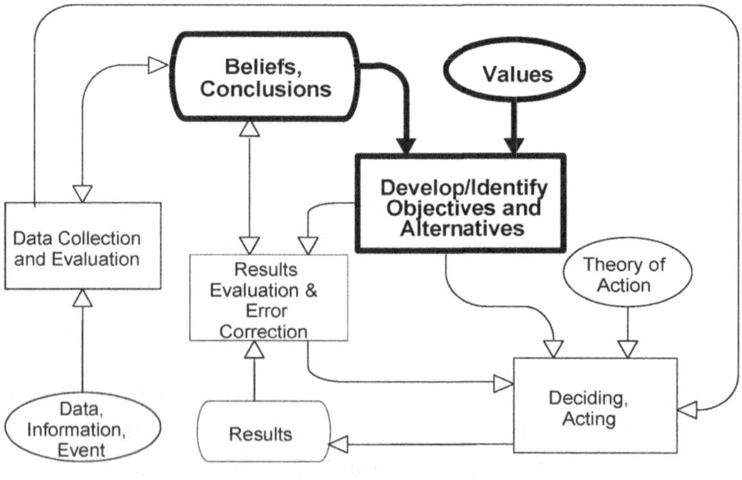

Thousands of books and innumerable magazine articles have been written about decision making. In the majority of them the subject is how to choose from amongst some given alternatives based on some given decision criteria or objectives. In real life decision problems the alternatives and objectives are not usually specified. In a real decision problem we usually must set the objectives and identify the alternatives. Little is written about how to do that. One exception is the work of Ralph Keeney on value-focused thinking (2). This chapter is largely based on his work.

Alternative-focused thinking

When faced with a decision problem our tendency is to first think of alternatives and, soon after having identified one or two, to begin evaluating them and even implementing one. We are action oriented; action is rewarded and there is often a need to show early "progress" toward a solution. Often little thought is given to explicitly identifying the objectives. The objectives may actually be developed after the fact as part of the report writing effort. As a result we frequently end up working the wrong problem for the wrong reasons, select an alternative that fails to meet an important objective identified too late, and fail to identify more effective alternatives.

Value-focused thinking

Why do you bother to make decisions? Why not just let whatever will happen, happen? It must be because the consequences of different alternatives matter to you. You make decisions because you care. You care because of your values. Your values are the reason you make decisions. Value-focused thinking begins with your values, and uses your values in a structured approach to generating better objectives and alternatives.

One Step or Two

You may consider setting objectives and identifying alternatives to be two separate steps. But alternatives and objectives are inextricably linked. It is impossible to think of objectives without alternatives popping into your head. And when you think of the features of an alternative it is common to recognize an item that should be an objective. There is symbiosis in dealing with objectives and alternative together.

Also, the boundary between objectives and alternatives is ambiguous. A well-developed objectives hierarchy should terminate naturally in the alternatives. For example, if you desire to buy a car and your objectives include: red, convertible, cheap; it is also fair to say that an alternative is to buy a cheap, red convertible.

Objective of this Chapter

To change your focus from alternative-focused thinking to value-focused thinking

3.2 EXAMPLES OF POOR OBJECTIVES IDENTIFICATION

Fire Truck

Keeney (2) tells a story about a community whose fire department took too long to respond to fires. A proposal to purchase more fire trucks was being debated and appeared headed for an election.

But an evaluation revealed inefficient use of existing equipment, such as too many trucks deployed to minor fires. More effective use of existing equipment solved the response-time problem.

The alternative of purchasing additional fire trucks was selected prematurely. It is a very salient and obvious alternative. A focus on the objectives revealed the better alternative.

Level Control Valve on Upper Level

A produced water treating system has a recycle line (for level control) from the discharge of the produced water pumps back to the produced water tank. Because the discharge of the pumps is several feet above the deck it made sense to the piping designers to continue the line up and put the level control valve on the upper deck. This works fine during normal operation. But during startup of the pump, if the line is not full of liquid, waterhammer may break it. The level control valve and the entire recycle line should have been located below the normal water level in the tank so that the line would always be full of water.

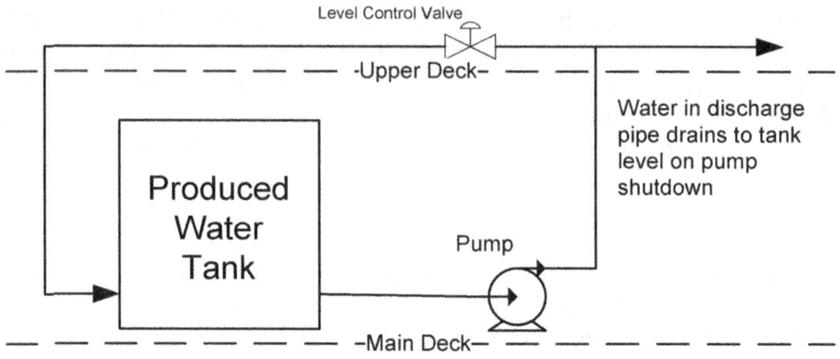

The piping designer had two objectives in mind, ease of pipe routing and ease of access to the valve, but was not aware of the very important objective of avoiding waterhammer on startup of the pump.

Waterflood Tank

A waterflood system features a water tank downstream of the filtration and deoxygenation systems and upstream of the high pressure injection pumps. At first glance the design makes sense. The tank provides a buffer between systems and provides a head of liquid for the injection charge pumps. It is very natural thing for a process engineer to put a tank there.

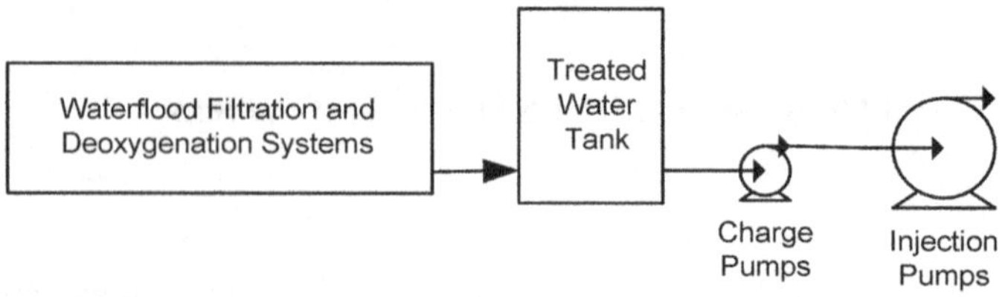

But the tank was not a good idea for several reasons including: a) the flowrate is such that it is not possible to provide enough volume for meaningful buffering between systems, b) the tank provides a place for oxygen ingress, c) the tank introduces significant complexity in the form of nitrogen blanketing and level control, d) head to the injection pumps could have easily been supplied by simply increasing the operating pressure of upstream equipment, and e) the tank adds significant weight to the platform.

The tank serves no significant useful purpose and causes significant problems. It was provided as an obvious solution to a typical problem with little thought given to the actual requirements in that situation.

Objectives Identified Too Late

The problem in all of the cases discussed above is selection of an alternative prior to effective identification of all of the important objectives. Our goal should be to identify all valid objectives and alternatives prior to making decisions. But what frequently happens is:

- An alternative is selected on the basis of an incomplete set of objectives.
- That alternative becomes the anchor point.
- The objectives that the chosen alternative meets come to be viewed as <u>the</u> valid objectives.
- Objectives identified late tend to be discounted if the chosen alternative is not effective at meeting them.
- Sunk costs of implementing the selected alternative bias later evaluations.
- Options decrease as time passes. At some point it becomes impossible to implement alternatives identified too late.

3.3 ADVANTAGES OF VALUE-FOCUSED THINKING

Value-focused thinking is a structured process that causes us to focus early on our real values when identifying objectives and alternatives.

There are many advantages to beginning with our values and using those values to guide development of objectives and alternatives including these from Keeney (2):

- **We are much more likely to find and select alternatives that satisfy our real needs.**

  Focusing on alternatives tends to anchor us on the first attractive alternative identified. Evaluation of that alternative will necessarily focus on the particular attributes of that alternative. The particular objectives satisfied by that alternative tend to become viewed as <u>the</u> valid objectives for solving the problem. We may fail to identify valid objectives that are not satisfied by the identified alternative.

  Conversely, early focus on objectives leads us to generate alternatives that are more directed towards satisfying those objectives.

- **In group settings, starting with values and objectives is a unifying process.**

  People will often agree on values and objectives even if they disagree on the alternatives. Starting a discussion by arguing about alternatives is likely to be divisive. Starting with a discussion of values and objectives gets everyone agreeing from the start rather than disagreeing. Agreeing on values and objectives makes it easier to work together on the alternatives.

- **Focusing on values allows participation of stakeholders who do not have the technical understanding to participate fully in discussion of alternatives**

  In highly complex endeavors like oil and gas development, choosing between design alternatives is a technical problem that only specialists can participate in. But non-technical stakeholders can be involved in identifying objectives.

- **Focusing on values allows coordinated treatment of problems featuring multiple decisions**

  When multiple related decisions need to be made, especially if they must be made by different people, early agreement on values and objectives provides a framework for coordination of the individual decisions.

3.4 OBJECTIVES STRUCTURING

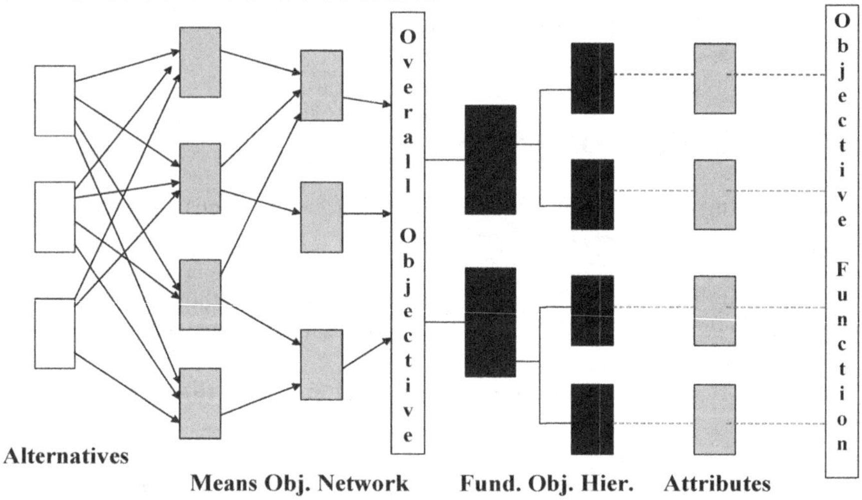

Structuring Objectives

The graphic above is an objectives structure. It summaries the nature of objectives. Objectives structures are useful for developing, understanding and communicating objectives.

At the center of the structure is the key objective or overall objective. The overall objective is the ultimate goal we wish to achieve. The overall objective divides the structure in two. There are two different types of objectives, fundamental objectives and means objectives.

Fundamental objectives are objectives that reflect what we really care about.

Means objectives are important only because they allow us to achieve fundamental objectives.

To determine whether an objective is a fundamental or a means objective, ask:

"Why do I care about this?"

If the answer is "because it allows me to achieve _____", it is a means objective.

If the answer is "because it's important", it is a fundamental objective.

The objective structure grows in two directions out from the overall objective.

Extending to the Right - Fundamental Objectives

To the right of the overall objective we have

- Fundamental objectives, the reasons we care.
- Attributes (measures of success).
- Objective Function.

Fundamental Objectives Hierarchies

Fundamental objectives can be structured into Fundamental Objectives Hierarchies. Not every fundamental objective has the same level of importance. For a given decision problem, one objective is likely to be the most important or overall fundamental objective. Other fundamental objectives will serve to further define or explain the overall objective.

Attributes (Measures of Success)

Since fundamental objectives measure what is really important to us we will want to assess how effectively any given alternative satisfies those objectives. Attributes are measures of success (see sections 3.9 to 3.12).

Objectives Function

To the right of the attributes is the objectives function. Multiple-objective decisions are often difficult because there are tradeoffs to make. Usually, no single alternative is best at satisfying all objectives. You must choose the alternative that, in some sense, is the best at satisfying all the objectives as a group. Usually some objectives are more important than others. Satisfying the most important objectives should take precedence over satisfying the lesser objectives. It may be useful to have a single measure of overall success. Such a single measure is called an objective function (see section 3.13).

Extending to the Left – Means Objectives

To the left of the overall objective lie the means objectives. Means objectives are ways of achieving fundamental objectives.

Means Objectives Networks

Means objectives can be organized into networks. Develop the means objectives network working from right to left. First identify the actions/objectives required to achieve the overall objective. Next identify actions required to achieve each of these means objective.

Alternatives

To the left of the means objectives network lie the alternatives. There is little difference between means objectives and alternatives. The means objective network terminates naturally in the alternatives.

Value Judgments => Fundamental Objectives;
Technical Expertise => Means Objectives

Note that value judgments are required for developing the Fundamental Objectives Hierarchy, but usually little technical expertise is required. Conversely, technical expertise is required to develop the Means Objectives Network. This suggests a division of labor – stakeholders who do not have the technical knowledge to participate in development of means objectives and alternatives may be able to participate in development of the fundamental objectives.

Details in Appendix A

Objectives structures are difficult and time consuming to build and are rarely used. They can be very useful for complex problems with multiple stakeholders. For more information on building them see Appendix A.

3.5 TECHNIQUES FOR IDENTIFYING OBJECTIVES

The de facto standard for generating lists of things is brainstorming. Brainstorming can be very effective, but it can also be limiting. There is a tendency in brainstorming to anchor on the first items mentioned and/or on the status quo. Guided brainstorming methods such as those discussed below can be more effective and efficient. Many of these methods are from Keeney (2). Another important source of objectives is lessons learned studies.

Lessons Learned Studies

After-action reviews and other lessons learned studies often identify objectives that were not obvious, or not made explicit, during the design phase.

Wish List

If there were no limitations what would you like to achieve?

Focus on the Uses - Develop Use Cases or Use Stories (see section 3.6)

Develop use cases or use stories if the decision involves an item or a process that people will use. A use story is a narrative describing how the item will be used. Develop a story for each possible use of the item. Detailed descriptions of system uses can identify objectives not readily apparent. Especially focus on abnormal situations such as startup, shutdown, upset conditions, routine maintenance, repair maintenance, operation with failed components, etc.

Use Alternatives

Which features of the identified alternatives are attractive? Which are not?

Identify a terrible alternative – What makes it terrible?

Identify a good alternative – What makes it good? What would you have to change to make an good alternative better?

For an already stated objective, ask which alternative would be ideal? Then ask what other implications that alternative would have?

Use a Different Perspective

List all the stakeholders and their interests. What do they want to achieve?

What would others want to achieve in this situation? What will your boss view as important? His boss? Your competitors? Your client?

View today from the future and the past. What will people 10 years from now or 100 years from now consider important? What would people 10 or 100 years ago have thought?

Assume a major obstacle has been overcome.

Assume a different status quo.

Structure Objectives (see sections 3.4 and Appendix A)

Create a fundamental objectives hierarchy and a means objectives network. The process of structuring objectives tends to reveal missing objectives.

Create Attributes (see sections 3.9 to 3.12)

Attributes are measures of success in meeting objectives. Determining measures of success can highlight what's important to you and what's missing. Creating attributes is especially important when multiple stakeholders exist.

Simply-stated objectives often have a "motherhood" quality to them that everyone agrees with. By defining how you intend to measure results, you more clearly specify the meaning of an objective. This will tend to surface areas of disagreement. If you can describe an objective in terms of a measurable attribute, then you truly understand it and others can understand it as well.

3.6 USE CASES

Use cases are an invention of the computer software business. Descriptions of the use case process can be found in several books on the subject including *Writing Effective Use Cases* (27) by Alistair Cockburn which is the source of the material in this section.

A use case is a description of how an item, system or process will be used to achieve particular goals. Each goal typically requires a different use case. Use cases are often nested with one use case referencing other use cases.

I find these very useful for specifying equipment items and systems. A description of how the equipment will be used (especially in abnormal situations) can avoid some embarrassing misunderstandings.

Use Case Contents:

Goal:	Each use case has a specific goal which the primary actor wants to accomplish.
Actors:	People, items, control systems, etc. that have goals and play roles.
Primary Actor:	Person with the specific goal that the use case is built to describe.
Stakeholders:	Everyone with a goal or interest in the outcome of the use case.
Triggering Event:	Event that starts the use defined by the use case.
Success Guarantee:	Definition of what needs to be accomplished for the lead actor to consider the use a success.
Minimum Guarantee:	The minimum that the system must accomplish if success is not achieved. For engineering designs this is usually a safe result such as a system shutdown to safe conditions.
Main Success Scenario:	A scenario which achieves success as the system is designed to operate (flawless operation).
Extensions:	A list of things that could go wrong with main scenario steps and what should happen when things go wrong (corrections, mitigations).

The two most important components of a use case are frequently the main success scenario and the extensions. The table below is the main success scenario for the goal of "Withdrawing cash from an ATM".

Example: Withdrawing Cash from Automatic Teller Machine (ATM)

> **Main Success Scenario:**
>
> 1) **Customer inserts card**
> 2) **Machine verifies card and asks for PIN number**
> 3) **Customer types PIN number and hits Enter Key**
> 4) **Machine verifies PIN and asks customer to select transaction type**
> 5) **Customer selects "Cash"**
> 6) **Machine asks customer to specify account to withdraw cash from**
> 7) **Customer selects account**
> 8) **Machine asks customer to specify the quantity of cash she desires**
> 9) **Customer specifies amount**
> 10) **System verifies amount acceptable and provides cash to customer**
> 11) **System asks if customer wants to make another transaction**
> 12) **Customer answers "no"**
> 13) **Machine prints receipt and returns customers card**

The main success scenario describes the successful operation of the system and hence contains directly the key objectives of the system for this particular use. It lists all necessary and expected steps if everything goes as planned.

But things often don't go as planned. Perhaps more important than the main success scenario are the extensions. Extensions are failure modes and alternative paths; anything which could cause a scenario different than the main success scenario. A few extensions to three of the steps in "Withdrawing cash from an ATM" are listed in the table below.

Extensions (Failure Modes)

2) System fails to read card
   a) inserted wrong
   b) invalid card
      1) expired
      2) withdrawn by bank
      3) invalid card (ie, wrong bank)
   c) Card damaged

3) PIN entry failure
   a) User inputs wrong PIN
      1) typing error
      2) forgot PIN
      3) card stolen – thief guessing
      4) PIN changed
   b) Keypad malfunction

10) Cash amount error/problem
   a) more than card limit
   b) more than in bank account
   c) more than in ATM machine
   d) not multiple of $10

To develop extensions, consider what can go wrong at each step in the success scenario and/or what other things might occur at each step. Possible failure modes identified via extensions suggest other system objectives.

3.7 EXERCISE – IDENTIFYING OBJECTIVES

Your boss has asked you to generate some ideas for training this year's class of new hires (engineers fresh out of school). You have to make a presentation to management this afternoon. Identify the objectives for this program using three methods described in the previous section:

Wish List,
Use Alternatives,
Use a Different Perspective

If you prefer substitute a decision problem of interest to you.

Decision Problem: ______________________________

Objectives:

3.8 EXERCISE – EXAMPLE RESULTS

Wish List:

Already Trained: It would be nice if the employees were already trained (maybe we should be hiring experienced engineers instead of fresh outs?)

Free: It would be great if we could train them without cost.
Some vendor training would be free.
Maybe they could train each other. (What skills do they have?)

From a Client's Perspective

Clients will be glad to see that we are training our new hires. Perhaps they'll help (enlightened self-interest). Maybe they have canned training programs we can send our new hires too. If the client also has new hires to train maybe we can invite client new-hires to our training and get some client relations points and sales benefit. Perhaps the Sales Department will share the cost of training.

Use Alternatives

In-House Seminars given by experienced staff

- Very time consuming for staff to develop training programs (objective = minimize time required of experienced staff).
- Can less experience people develop the training? This would provide training for them as well. Perhaps the new-hires could generate some of the training material themselves under the supervision of mentors.
- We can develop training material once and use it every year (objectives = reuse what we can in the future and use what we can from the past).
- Opportunity for experienced staff to get what they know on paper in the form of process manuals, go-by's, etc.
- Maybe we could market these seminars after they've been developed and used in-house.

Public courses

- Expensive.
- Opportunity to meet other people, network.
- Minimizes staff commitment – no training to develop.
- Maybe we can send each new hire to a different course and have each teach the others what they learned (have an experience engineer sit in on sessions to correct errors and answer questions).

On-the-job training

- Best training, but slow. Some training needs to be accomplished quickly in classrooms.
- Can we find appropriate on-the-job training opportunities for all new hires.
- Provides tacit knowledge (things we know, but don't know we know).

3.9 ATTRIBUTES – MEASURES OF SUCCESS

When we evaluate alternatives we try to determine which available alternative best fulfills our objectives. Hence, we should try to measure the degree to which each important objective will be met by each of the various alternatives. We can do this by assigning attributes, or measurement scales, to each of the lowest level fundamental objectives.

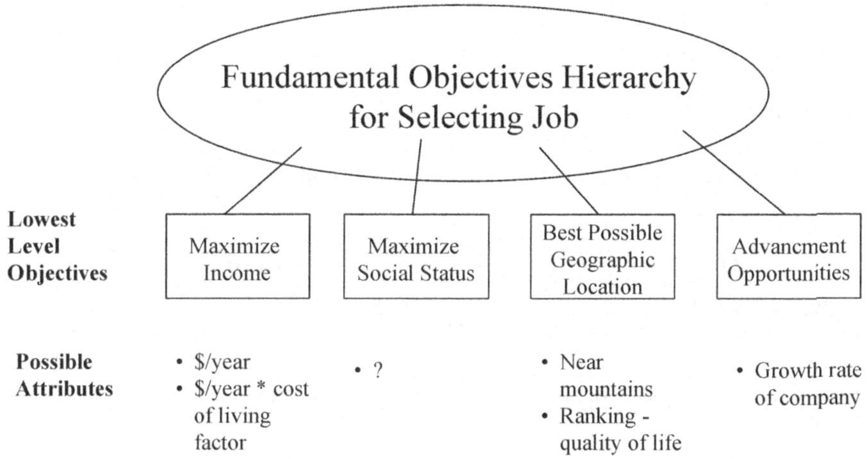

The graphic above shows a simple example. In trying to select between job offers in different cities, we might have a fundamental objectives hierarchy which ends in the lowest level objectives:

- Maximize income
- Maximize social status
- Best possible geographic location
- Maximize advancement opportunities

This seems like a reasonable set of objectives. But the objectives are not defined well enough to provide clear direction for decision making. For instance, what does it mean to maximize income? Do you mean gross income? After tax income? Disposable income (what's left after the necessary bills are paid)? What do you plan to do with the extra income? If you would use disposable income mainly to go on ski vacations, then a lower salary in Salt Lake City (near ski resorts) might be more lucrative than a significantly higher salary in Atlanta.

This illustration of the variety of possible meanings for such a simple objective as "maximize income" suggests the best reason for developing attributes: to help you and others truly understand what your objectives are. Simply-stated objectives often have a "motherhood" quality about them that everyone agrees with. By defining how you intend to measure results, you more clearly specify the meaning of an objective. This will tend to surface areas of disagreement. If you can describe an objective in terms of a measurable attribute, then you truly understand it; and others can understand it as well.

3.10 TYPES OF ATTRIBUTES

It is not always easy to define measurable attributes. If your true objective for maximizing income is to get the largest before tax salary, then that attribute is salary in dollars, a very easy measure. On the other hand, coming up with an attribute for "maximize social status" will be much trickier. We can group attributes into three categories:

- Natural attributes
- Constructed attributes
- Proxy attributes

Natural Attributes

Natural attributes are those which are easily identified with the objective such as:

Objective	Attribute
Maximize income	Salary in dollars
Minimize auto accidents	Number of accidents reported to police
Maximize sales of game programs	Number of programs sold

Constructed Attributes

When you rank something on a scale of 1 – 10 you are constructing an attribute scale.

How would you measure success for the objective "maximize social status"? There is no natural scale. But with a little creativity you may be able to construct an appropriate scale.

Proxy Attributes

Sometimes it is possible to measure something that is statistically correlated to the objective we are really interested in.

Proxy attributes are very popular. The grades of students on standardized tests might be a reasonable proxy for the effectiveness of the school they attend (or might not, depending on circumstances). The unemployment rate is used as a proxy attribute for the health of the economy. If you are shopping for a car and safety is a very important objective, you might use the weight of cars as a proxy attribute for safety – larger, heavier cars tend to survive crashes better than smaller, lighter ones.

3.11 CONSTRUCTED ATTRIBUTES

A constructed attribute is a scale used for ranking an alternative against specified standards. For example, suppose we wish to measure the competence of engineers and one of the areas we wish to measure is their familiarity with applicable industry codes. The table below is a constructed scale for measuring the proficiency of engineers with codes on a scale from 0 to 5.

Measure	Engineer Competence Level
0	Entry level engineer. No useable knowledge of code content
1	General familiarity with the codes important to his/her field. Cursory working knowledge of the key codes. Can apply specific codes rules with guidance of superior.
2	Working knowledge of key codes in his/her field. Understands when particular codes apply. Can apply key codes without supervision.
3	Working knowledge of all commonly applied codes in his/her field. In depth knowledge of key codes.
4	Recognized within the company for expertise in one or more codes. Resource for others in interpreting meaning of codes and/or conflicts between codes.
5	Recognized within the industry as an expert in one or more codes. Provides expertise to the industry in observable way, such as publications, presentations at conferences and membership on code committees.

The rating scale is the familiar 0 to 5 rating scale. Much of the arbitrariness usually associated with such scales is removed by describing what is meant by each rating.

Constructed scales can be complex using multiple criteria as shown in the graphic below. In this scale the adequacy of a design (0 to 3) is based on both the potential consequence (or severity) of an accident and the likely frequency of it occurring. For example, an event that has negligible potential to cause injury is acceptable at any frequency. On the other hand, an event likely to cause a single fatality every 10 years calls for design improvement.

Frequency (once per)	Potential Harm				
	Many Fatalities	Single Fatality	Serious Injury	Minor Injury	No Injury
Month	3	3	3	1	0
Year	3	3	2	0	0
Decade	3	2	1	0	0
100 years	2	1	0	0	0
1000 years	1	0	0	0	0

Where:

3: Unacceptable
2: Design revision or management controls mandatory
1: Consider design improvement and/or mitigating action
0: Acceptable as is

3.12 PROXY ATTRIBUTES

The idea of a proxy attribute is to substitute something you can measure for something you cannot easily measure. Of course, you should have a good reason to believe that there is a correlation between the two.

The Dow Jones Industrial Average is a well-known proxy attribute. When it was developed (before powerful computers) it would have been impossible to calculate how the NYSE was faring at any given moment. To avoid the computational problem, an average of 30 leading stocks was developed as an indicator. That measure is still used today. It is a proxy for the condition of the market as a whole. It is tacitly assumed that the market as a whole behaves pretty much as those 30 behave.

If you want to measure the effectiveness of a teacher, give her students a test. The average grade of her students is a proxy measure of her skill as a teacher. It's not necessarily a good measure because many other factors also play a role in success of the students.

3.13 OBJECTIVE FUNCTION

Multiple-objective decisions are often difficult because there are tradeoffs to make. Usually, no single alternative is best at satisfying all objectives. You must choose the alternative that, in some sense, is the best at satisfying all the objectives as a group. Usually some objectives are more important than others. Satisfying the most important objectives should take precedence over satisfying the lesser objectives.

It may be useful to have a single overall measure of success. Such a single measure is called an objective function.

Words of Caution

First, a critique of objective functions in general. In order to build an objective function you will have to make judgments of the relative importance of each objective and possibly of the relationships between individual objectives. Once this is done, the overall decision becomes easy – you simply select the alternative with the highest objective function value.

But:

- What reason is there to believe that you will be any better at making the judgments required to construct the objective function than you would be at making a single decision between alternatives.
- The relationships between objectives are usually more complicated than that captured in an objective function. Objective functions simplify the problem and may lead to inappropriate decisions, especially at the extremes.

Reasons for Developing Objectives Functions

Despite these significant limitations, there are often very good reasons for developing objective functions. These include:

- When the objectives are all monetary (no value judgments are required).

- When the decision must be made by someone without detailed knowledge of the problem (the objective function provides expert guidance).
- When the decision must be auditable or when you want to avoid appearance of bias in the decision.
- Developing an objective function forces you to evaluate your preferences and articulate them clearly. It can be a valuable exercise even if the objective function is not directly used for making the decision.

An example of an appropriate use is for making competitive bid evaluations. An objective function allows you to structure the decision ahead of time. If done well you can accurately account for important factors other than price and make an intelligent decision without the appearance of bias or favoritism.

Weighted Objective Function

If attributes are defined for every lowest level fundamental objective, they can be combined into an objective function.

Typical Objective Function (OF):

Objectives Independent

$$\textbf{Value} = W_1 * X_1 + W_2 * X_2 + \ldots\ldots + W_n * X_n$$

Linear

Relationships are usually more complicated than this. It is very difficult to get a value model that accurately reflects the true objectives.

The most commonly encountered general-purpose objective function is a linear, weighted function as shown in the box above. The model is of the form:

- Rate every objective from 1 – 10 (or other similar scale), **Xi's**
- Assign a weight to each objective, **Wi's**
- Sum the weighted attributes

Yielding: $\text{Value} = W_1*X_1 + W_2*X_2 + \ldots\ldots + W_n*X_n$

This formulation assumes that all objectives are independent and that our preferences are linear with respect to each objective.

Let's see how that could get us in trouble. Suppose that you intend to buy an accounting program (computer software), hire a programmer to learn the software and configure an application for your company, and train your staff to use it. Other than cost, your objective function for this decision includes three equally weighted attributes 1) ease of programming/configuring, 2) technical merit of the software, and 3) ease of use (by end users).

The three packages are rated thusly (scale of 1 to 10):

Package	Ease of Prog	Tech Merit	Ease of Use	Sum
A	10	8	5	23
B	4	9	9	22
C	7	7	7	21

Package A has the highest rating and C has the lowest. But, this is likely to be the reverse order of your real preference in this case. Option A saddles your staff with a hard to use program, Option B saddles your programmer with a difficult task (potentially resulting in late delivery, poor implementation, and difficult, expensive maintenance). Option C is probably the best choice. That's easy to see now, but how could you have captured that wisdom a priori. If this were a government project you might now be stuck with an inferior choice, though in practice that probably wouldn't happen. Faced with this situation you would probably just adjust your ratings to yield the "right" answer. These techniques are probably used more to justify decisions than to make them.

3.14 USING OBJECTIVES TO IDENTIFY ALTERNATIVES

We now turn our thoughts to the identification of alternatives. We want to identify excellent alternatives which accomplish all of our objectives. Now that we really know what our objectives are we have a chance of accomplishing that and we should use those objectives to guide the search.

The most visible, and so most available, alternative for many decision problems is the status quo. We tend to be anchored on the status quo. Alternatives we identify tend to be movements away from the status quo. Identification of liberating alternatives sometimes requires that we pull up the anchor and set it down somewhere else. The place you'd like to be anchored is at the ideal solution. Our goal is to identify the alternatives that best meet our objectives. Several methods are suggested below (largely from Keeney (2)).

Consider one objective at a time:

A good way to begin is to think of one objective at a time and try to identify alternatives that are very good at meeting that one objective. Many of the alternatives generated this way will fare poorly at meeting other objectives. Indeed, this should be the result. If you've been creative at generating alternatives that are ideal in some way, then some of them will surely be awful in other ways.

Try to modify an alternative in order to accommodate other objectives

An alternative specifically designed to satisfy one particular objective may be poor at meeting other objectives. Study the better alternatives and try to change them in ways that better meet the other objectives without sacrificing too much.

Try to combine alternatives:

After generating a list of alternatives for individual objectives, consider whether two or more can be combined.

Can two or more alternatives be combined into a single solution?

Can two or more alternatives be implemented independently to yield the desired effect?

Broaden the Decision Context

It is likely that the decision context is limited to things you can control and influence. In identifying alternatives we should give some thought to what alternatives would exist if we broadened the decision context. Are there alternatives available to higher levels of management, or to other parallel groups, that are better than any you could implement?

Focus Thinking on Desirable Consequences

Our goal is to avoid anchoring on the status quo. Consider dropping anchor at the ideal solution instead. Imagine that every attribute is maxed out; you have perfectly or ideally satisfied every objective. What sorts of alternatives would achieve that. If you cannot identify any alternatives that gets you there, relax one or more attributes slightly. What is the minimal compromise you can make before arriving at a feasible alternative?

Use Analogies from Other Fields - TRIZ

Often a problem solution in another field can yield ideas. Genrich Altschuller, a Russian Mechanical Engineer and patent expert, studied over 40,000 patents and developed a structured method for guiding creativity. Altschuller began with the observation that most patents resulted from a creative solution to a conflict between objectives. For example: an airplane needs to be light in order to fly and strong in order to withstand the forces it is subjected to. Adding steel makes it stronger, but heavier. Altschuller identified many types of conflicts and catalogued the approaches people used to solve each type of conflict. The result is a structured guide for creative problem solving.

The method is called TRIZ (Russian acronym for Theory of Inventive Problem Solving).

<u>TRIZ Example</u> (from website www.mazur.net/triz/):

Aluminum beverage cans should be light (thin walled) and should also be able to support weight when they are stacked on top of each other. These two requirements are in conflict. TRIZ suggests the following possible solutions to the conflict:

- Segmentation (segment item)
- Spheroidality (replace linear parts with curved ones)
- Transformation of chemical or physical states of an object

Look at a soda can. The sides are not cylindrical all the way up, but curve inward near the top and bottom – an application of spheroidality.

3.15 INFLUENCE DIAGRAMS

A good summary tool for conveying objectives to others, and for insuring completeness in your planning, is the influence diagram.

As the name implies, influence diagrams show the influence of individual items in a decision model.

Symbols:

Rectangles indicate decisions.

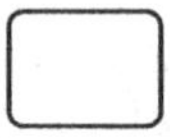

Rounded corner boxes indicate consequences.

Ovals indicate chance events or givens. Things we have no control over such as weather, actual oil reservoir size, die toss result, future cost of energy, etc. are shown as chance events.

Arrows indicate direction of influence.

It is important to note that arrows do not indicate sequence or chronology. The influence diagram shows influences at a point in time, not sequences over time.

Uses of Influence Diagrams

Solving the Decision Problem:
If all the influences can be defined and all the input data quantified, the influence diagram can be solved.

Insuring Completeness:
Influence diagrams are very useful for ensuring completeness. I use them to help identify and account for everything that has impact on the problem I'm working on.

Managing Interfaces: Where separate work teams are working on aspects of the same problem, an influence diagram identifies important interface areas

Gas Buy-back Influence Diagram

Below is an example of an influence diagram that I used to plan a gas buy-back study. Oil production platforms frequently backflow gas (buy-back) through the Sales Gas Line to purge and pressurize the facility and provide fuel prior to startup. In this case the initial gas buy-back event presented several potential problems including: 1) low temperature (due to a large pressure drop) with resultant metallurgy specification issues, 2) liquid dropout, including water with resultant hydrate (freezing) potential, and 3) high velocities in two-phase streams resulting in erosion potential. The influence diagram effectively summarizes a relatively complex problem.

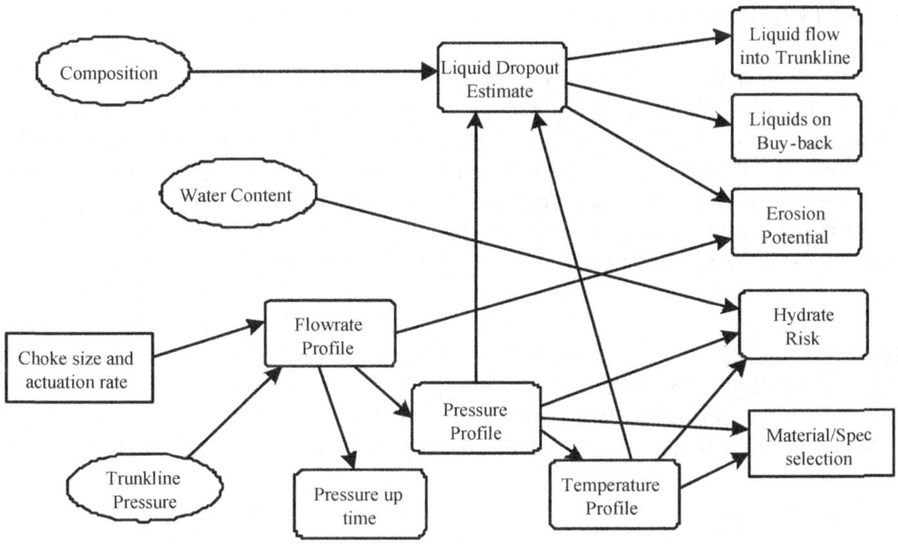

The diagram shows two decision points:

Choke Size and Rate:	The choke (valve used to control the flowrate during buy-back) must be sized and the rate of opening specified.
Material Specification:	The resulting pressure profile and temperature profile provide requirements for piping material specification.

There are three givens or variables subject to chance:

Gas composition (hydrocarbons, H_2S, CO_2)
Water content of the gas
Trunkline operating pressure

All other aspects of the problem are results (calculations).

Suggestions:

- Don't get hung up on node shape. I often use only ovals or rectangles. It is usually clear from the context which nodes are decisions and which are consequences.
- Focus on completeness.
- When using the diagram as a coordination tool, make sure everyone interprets each node the same. The short block title can easily be misinterpreted.

3.16 CHAPTER SUMMARY

When faced with a decision problem our tendency is to first think of alternatives and, soon after having identified a couple, to begin evaluating them and even implementing one. We are action oriented; action is rewarded and there is often a need to show early "progress" toward solution. Often little thought is given to setting the objectives. The objectives may actually be developed after the fact as part of the report writing effort. As a result we often work the wrong problem for the wrong reasons, select an alternative that fails to meet an important objective identified too late, and fail to identify more effective alternatives.

Value-focused thinking is a structured process that causes us to focus early on our real values. There are many advantages to beginning with our values and using those values to guide development of objectives and alternatives including:

- We are much more likely to find and select alternatives that satisfy our real needs.
- In group settings, starting with values and objectives is a unifying process.
- Focusing on values allows participation of stakeholders who do not have the technical understanding to participate fully in discussion of alternatives
- Focusing on values allows coordinated treatment of problems featuring multiple decisions

Several techniques for identifying objectives were presented. One effective method is developing use cases. Use cases are descriptions of how an item, system or process will be used to achieve particular goals.

Objectives can be structured in Fundamental Objectives Hierarchies and Means Objectives Networks. Objectives structures are useful for developing, understanding and communicating objectives.

Fundamental objectives are objectives that reflect what we really care about.

Means objectives are important only because they allow us to achieve fundamental objectives.

Effective identification of the objectives facilitates the identification and evaluation of alternatives. When we evaluate alternatives we try to determine which available alternative best fulfills our objectives. Hence, we should try to measure the degree to which each important objective will be met by each of the various alternatives. We can do this by assigning attributes, or measurement scales, to each of the lowest level fundamental objectives. We can group attributes into three categories:

- Natural attributes
- Constructed attributes
- Proxy attributes

If all the fundamental objectives can be measured it may be possible to generate an objectives function that provides one measure of success useful for comparing the alternatives.

3.17 ANSWERING THE FOUR QUESTIONS

What observations from this chapter can we make about our four key engineering questions?

Why do so many changes occur late in projects?

Objective(s) identified late:

I believe that the number one cause of late changes is the failure to adequately identify design objectives early in the design effort. Objectives identified late are often not achieved by the selected design. Often these issues are raised when someone new joins the project. For instance, when the operations representatives are finally assigned they may quickly find many things they disagree with. Design engineers are usually not sensitive to all the needs (objectives) of operators and other stakeholders. We tend to generate designs that meet the obvious objectives and then to gradually identify other objectives over the course of the design.

Objectives not fully defined or understood:

Simply stated objectives can have a motherhood quality; everyone thinks they understand. But we frequently misunderstand and we disagree when we finally realize that what's being built isn't what we anticipated.

Attractive alternative identified late:

Some good ideas naturally occur late in the design process as more and more people become aware of the design details. We all want to do the right thing and the best thing. Hence it is difficult to ignore a bright idea even if it occurs very late in the game.

Why do we have so many problems at interfaces?

Objectives not fully defined or understood:

This is especially prevalent across team interfaces. Objectives not full defined are easily misinterpreted by separate teams.

Different teams develop incompatible solutions:

There are literally an infinite number of ways to design any project. It is possible for two teams to meet all the identified objectives, but to generate designs which are incompatible.

Conflicting objectives:

Different teams will naturally have different design objectives. If those objectives conflict (impossible to achieve all objectives simultaneously) there will be interface conflicts.

Teams disagree on the best alternative:

Even if all teams understand the objectives and work toward meeting them, they can easily disagree on which alternative best achieves the objectives.

Why do we repeat past mistakes?

Alternative-focused thinking:

Alternative-focused thinking tends to anchor us to the alternatives we've applied in the past.

Why do we so often finish behind schedule?

<u>Trying to satisfy unrealistic objectives:</u>

Engineers writing specifications tend to ask for the 'moon' without thinking through the consequences. Other engineers trying to achieve the specification then spend inordinate amounts of time trying to achieve the impossible.

<u>Evaluating too many alternatives:</u>

Evaluating too many alternatives for too long costs time and money and distracts attention from the real issues.

Chapter 4
Deciding

Sources:

Utility theory was developed by von Neumann and Morgenstern (25)

Prospect theory was developed by Amos Tversky and Daniel Kahneman (4, P1, P2)

The discussion of regret is largely from a remarkable article by Gilovich and Medvec (P3)

The discussion of decision strategies at the end of the chapter comes largely from Gigerenzer (14)

Preface:

This chapter discusses techniques useful for selecting the preferred alternative given a set of objectives. This is what most people mean by the words "decision making". Most of the research done on decision making falls within the realm of this chapter. There is an incredibly large body of work to choose from.

This chapter is obviously a required chapter in a course about decision making, but it is the least important chapter in this book. For most decision problems, other aspects of the problem such as believing the right things, setting the right objectives, identifying suitable alternatives, and acting in effective ways are much more important.

Do I want you because you're wonderful, or are you wonderful because I want you?

From: Do I Love You Because You're Beautiful
Rogers and Hammerstein

CHAPTER 4 CONTENTS

4.1 INTRODUCTION
4.2 PROBABILITY AND STATISTICS - EXERCISE
4.3 EXERCISE ANSWERS
4.4 BAYESIAN REASONING
4.5 HOT STREAKS AND COINCIDENCES
4.6 THE BELL CURVE - NORMAL PROBABILITY DISTRIBUTION
4.7 EXERCISE – NORMAL DISTIBUTIONS
4.8 NORMAL DISTRIBUTION EXERCISE RESULTS
4.9 THINKING IN TERMS OF DISTIBUTIONS
4.10 NORMAL DISTRIBUTION CALCULATION VIA EXCEL™
4.11 SENSITIVITY – TORNADO DIAGRAM
4.12 MONTE CARLO SIMULATION
4.13 OPTIMIZATION
4.14 LINEAR PROGRAMMING EXAMPLE - OIL WELL PRODUCTION
4.15 LINEAR PROGRAMMING VIA EXCEL
4.16 DECISION TREES
4.17 MODELING PREFERENCES
4.18 VALUE CURVE
4.19 PROSPECT THEORY – REFERENCE EFFECT FOR RISK TOLERANCE
4.20 REGRET
4.21 REGRET THEORY
4.22 MODELING PROSPECT THEORY OBSERVATIONS VIA REGRET THEORY
4.23 SUMMARY – PREFERENCE MODELING
4.24 DECISION STRATEGY
4.25 CLASSES OF DECISION STRATEGY
4.26 PARADOX – LESS IS MORE
4.27 WORKPLACE DECISION STRATEGIES
4.28 CHAPTER SUMMARY
4.29 ANSWERING THE FOUR QUESTION

4.1 INTRODUCTION

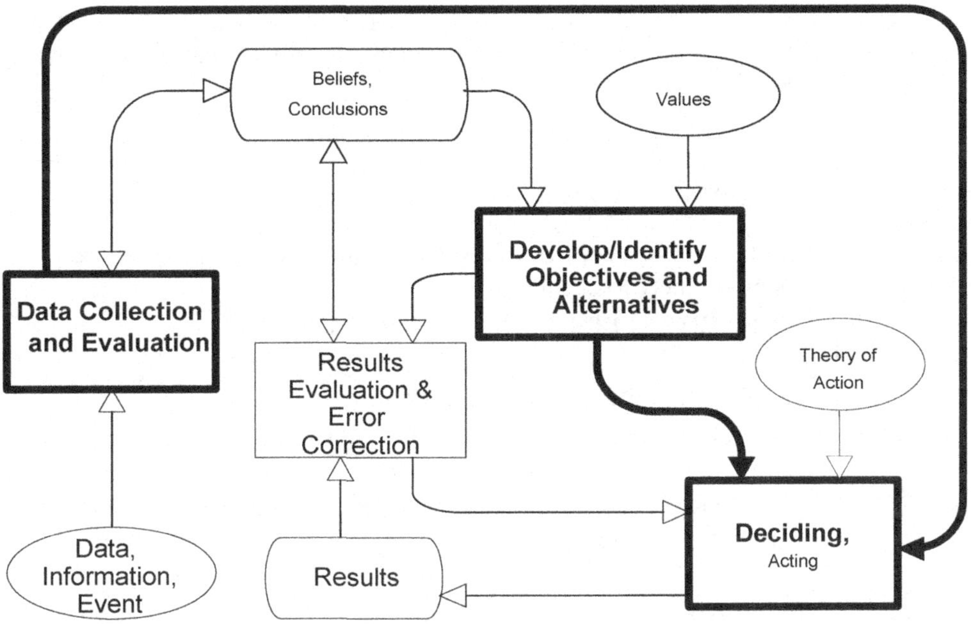

The graphic above covers the task of deciding, the process of selecting one of the identified alternatives. This is what most people mean when they say 'decision making." The node in which "deciding" resides on the diagram also includes "acting" - the process of taking action to implement the decision. You may question why "deciding" and "acting" are included in the same node. They seem like separate steps. Whether deciding and acting are separate steps is partly semantics; *is the decision made when a commitment to act is articulated, or when the act is performed?*. But this is much more than a semantics argument.

Many otherwise excellent decisions never get effectively implemented. Since it is actions (and subsequent results) which are important to us, we should be conscious of selecting alternatives which can and will be implemented. Chapter 5 covers the subject of acting.

This chapter contains a brief introduction to deciding. More information is contained in Appendices B, C and D. Topics covered include:

- Probability and statistics
- Bayesian reasoning
- Optimization
- Preference Modeling (Utility Theory, Prospect Theory, Regret Theory)
- Decision strategies

4.2 PROBABILITY AND STATISTICS - EXERCISE

Making selections requires evaluation of data. The fields of probability and statistics address methods of evaluating the meaning of data. Hopefully you took some statistics in college or picked it up elsewhere.

Exercise:

1. If you flip a coin 10 times, which of the following sequences is more likely to occur?

a. HTHHTHTTTH b. HHHHHHHHHH

2. Which of these is more likely?

a. HTHHTHTTTH b. HHHHHTTTTT

3. A rare disease has been found in your neighborhood. Health officials estimate that 1 in 1000 people in your area have the disease. A test is available. The test is 99% accurate (if you are sick there is a 99% probability the test will be positive, if you are well there is a 99% probability the test will be negative). If you test positive, what are the odds that you are sick?

4.3 EXERCISE ANSWERS

Question 1:

We can work out the problem mathematically. Starting with question 1a: The odds of getting H on the first flip is 0.5 (50%), odds of getting T on the second flip is also 0.5, odds of H first followed by T is 0.5*0.5 = 0.25, odds of getting H next is also 0.5 and so the odds of the first three being HTH are 0.5*0.5*0.5 = 0.125. Continuing this pattern through ten flips yields a probability of getting answer a,
P(a) = .5*.5*.5*.5*.5*.5* .5*.5*.5*.5 = 0.000997 or 1 in 1024

By the identical argument, the probability of getting answer b, ten H's in a row is, P(b) = 0.000997 or 1 in 1024.

Each sequence is equally likely to occur.

Question 2: Again, by the arguments above, each listed sequence has a probability of occurring of 0.000997 or 1 in 1024.

Discussion of Questions 1 and 2:

Why do the "a" answers seem more likely to the casual observer?

In question 1, answer "a" seems more likely because we expect that in ten flips we will get about 5 heads and about 5 tails. It is true that we are more likely to get 5 heads than 10 heads, but that's because there are many sequences that yield 5 heads and only one sequence that yields 10 heads. The probability of any particular one of the sequences yielding 5 heads occurring is the same as the probability of the one sequence yielding 10 heads occurring.

Similarly, in question 2, there are many sequences yielding 5 heads, but only one sequence in which the heads are grouped at the beginning as in "b".

Question 3

This one is trickier. The answer is 9%. If you answered 99% you answered the wrong question. The question is "If you test positive what are the odds that you are sick?" The question you answered is "If you are sick what are the odds that the test will be positive?"

4.4 BAYESIAN REASONING

Question 3 above can be answered via application of Bayes theory.

$$P(B/A) = \frac{P(A/B)P(B)}{P(A/B)\,P(B) + P(A/\overline{B})P(B)}$$

$P(B)$ = Probability of B occurring
$P(A)$ = Probability of A occurring
$P(A/B)$ = Probability if A occurring if B occurs
$P(B/A)$ = Probability if B occurring if A occurs
$P(A/\overline{B})$ = Probability of A occurring if B does not occur

The problem is solved in Appendix B per the above equation. To apply Bayes Theory in an intuitive way:

- Suppose that 1000 people are tested the day that you are.
- Odds are about 1 person in that group is sick. The one ill person will probably test positive (99% chance)
- Odds are that 999 are well. But 1% of them will test positive (false positives). That amounts to 999 (.01) = 10 false positives.
- Hence, out of a group of 1000, of whom only about one is sick, about 11 will test positive.
- If you test positive there is a 9% chance that you are sick and a 91% probability that your test result is a false positive.

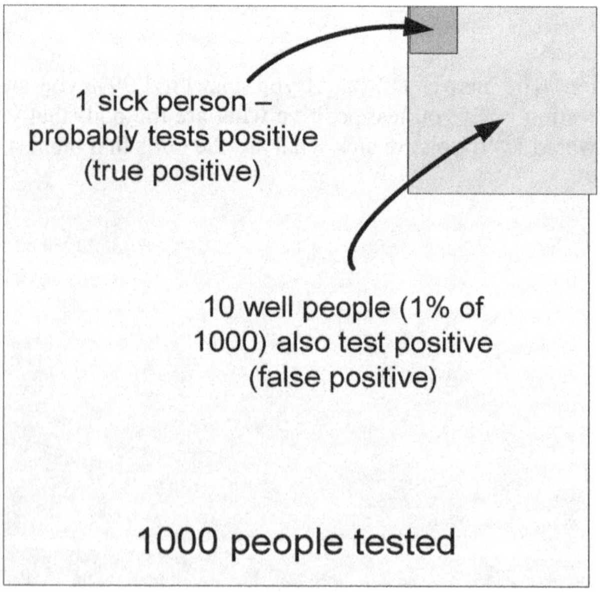

What if you belong to a high risk group?

Suppose that the population can be divided into two groups; one group with low risk (1 in 1000) and one group with very high risk (1 in 10). We saw above that if you belong to the low risk group that a positive test suggests a 9% chance that you are sick. What if you belong to the high risk group and have a positive test?

If 1000 high risk people are tested we would expect:

100 sick people, 99 of whom will test positive

and 900 well people, 9 of whom will test positive

If you are in the high risk group and you test positive your odds of being sick are about 99/108 = 92%

Example – AIDS Testing

Testing for AIDS (HIV infection) is a good example. Despite a very accurate test about 50% of people with no risk factors who test positive for AIDS are <u>not</u> infected. (from *Adaptive Thinking*, Gigerenzer).

Bayes theory provides a rational approach for incorporating new information into your beliefs. In the examples above, most people who test positive immediately assume that they are sick; a Bayesian thinker is concerned, but realizes that this new piece of evidence is far from conclusive.

Example: Oregon Attorney Charged with Aiding Al Qaeda in Madrid Bombings

On 11 March 2004 terrorists set off several bombs on commuter trains in Madrid, Spain killing nearly 200 people. Two months later the FBI arrested Brandon Mayfield, a American lawyer living in Oregon. The FBI identified Mayfield as a suspect based on fingerprints on a bag of detonators. Except for the fingerprints there was no reason to believe that Mayfield had any connection to the terrorists. He had never even traveled outside the US.

Two weeks later Spanish authorities matched the fingerprints to Ouhnane Daoud, an Algerian, and issued an arrest warrant. The FBI admitted that they had erred in the fingerprint analysis and Mayfield was released..

A Bayesian reasoner likely would not have made this mistake. Given the lack of a credible connection between Mayfield and the terrorists, a Bayesian reasoner would have assumed that the finger print match was a false positive.

4.5 HOT STREAKS AND COINCIDENCES

At Bat	Random	Hit?	Avg	At Bat	Random	Hit?	Avg	At Bat	Random	Hit?	Avg
1	0.287	1	1.000	33	0.61	0	0.394	65	0.931	0	0.308
2	0.994	0	0.500	34	0.112	1	0.412	66	0.661	0	0.303
3	0.432	0	0.333	35	0.604	0	0.400	67	0.836	0	0.299
4	0.765	0	0.250	36	0.876	0	0.389	68	0.093	1	0.309
5	0.157	1	0.400	37	0.972	0	0.378	69	0.823	0	0.304
6	0.959	0	0.333	38	0.547	0	0.368	70	0.176	1	0.314
7	0.901	0	0.286	39	0.831	0	0.359	71	0.752	0	0.310
8	0.416	0	0.250	40	0.463	0	0.350	72	0.299	1	0.319
9	0.388	0	0.222	41	0.822	0	0.341	73	0.441	0	0.315
10	0.964	0	0.200	42	0.979	0	0.333	74	0.651	0	0.311
11	0.496	0	0.182	43	0.992	0	0.326	75	0.503	0	0.307
12	0.231	1	0.250	44	0.501	0	0.318	76	0.528	0	0.303
13	0.226	1	0.308	45	0.942	0	0.311	77	0.849	0	0.299
14	0.600	0	0.286	46	0.508	0	0.304	78	0.616	0	0.295
15	0.974	0	0.267	47	0.687	0	0.298	79	0.277	1	0.304
16	0.137	1	0.313	48	0.449	0	0.292	80	0.244	1	0.313
17	0.900	0	0.294	49	0.159	1	0.306	81	0.174	1	0.321
18	0.286	1	0.333	50	0.559	0	0.300	82	0.059	1	0.329
19	0.231	1	0.368	51	0.833	0	0.294	83	0.666	0	0.325
20	0.928	0	0.350	52	0.992	0	0.288	84	0.137	1	0.333
21	0.009	1	0.381	53	0.968	0	0.283	85	0.679	0	0.329
22	0.515	0	0.364	54	0.466	0	0.278	86	0.602	0	0.326
23	0.556	0	0.348	55	0.174	1	0.291	87	0.053	1	0.333
24	0.822	0	0.333	56	0.715	0	0.286	88	0.489	0	0.330
25	0.025	1	0.360	57	0.242	1	0.298	89	0.735	0	0.326
26	0.499	0	0.346	58	0.777	0	0.293	90	0.666	0	0.322
27	0.072	1	0.370	59	0.002	1	0.305	91	0.288	1	0.330
28	0.041	1	0.393	60	0.152	1	0.317	92	0.843	0	0.326
29	0.287	1	0.414	61	0.852	0	0.311	93	0.154	1	0.333
30	0.466	0	0.400	62	0.052	1	0.323	94	0.816	0	0.330
31	0.148	1	0.419	63	0.954	0	0.317	95	0.435	0	0.326
32	0.414	0	0.406	64	0.414	0	0.313	96	0.099	1	0.333

Our lack of understanding of probability can easily lead us to erroneous conclusions. Above is the simulated batting record of a good hitter (baseball/softball) over a 24 game period in which she batted 4 times per game for a total of 96 at bats. The result of each at bat (a hit or not a hit) was determined via a random number generator, designed to yield a hit 1/3 of the time on average. Over the entire period, our player's batting average came out exactly as would be predicted, .333 (one hit in every 3 at bats).

But, over the course of the season, this random behavior resulted in one major slump (1 hit in 20 at bats in at bats 35 thru 54) and a major hot streak (11 hits in 20 at bats in at bats 12 thru 31).

This prompts the question: are ballplayer slumps and streaks caused by the actions of the player, or are they simple statistical phenomena over which the player has little or no control?

4.6 THE BELL CURVE - NORMAL PROBABILITY DISTRIBUTION

We have discussed so far only discrete events. Many events can yield a range of possible values. Many of these can be adequately described via a normal probability distribution. For such populations, the mean, μ, and standard deviation, σ, provide a wealth of insight.

The mean is the average value.

The standard deviation, σ, can be thought of as a best guess of how far the outcome of a typical experiment might lie from the mean.

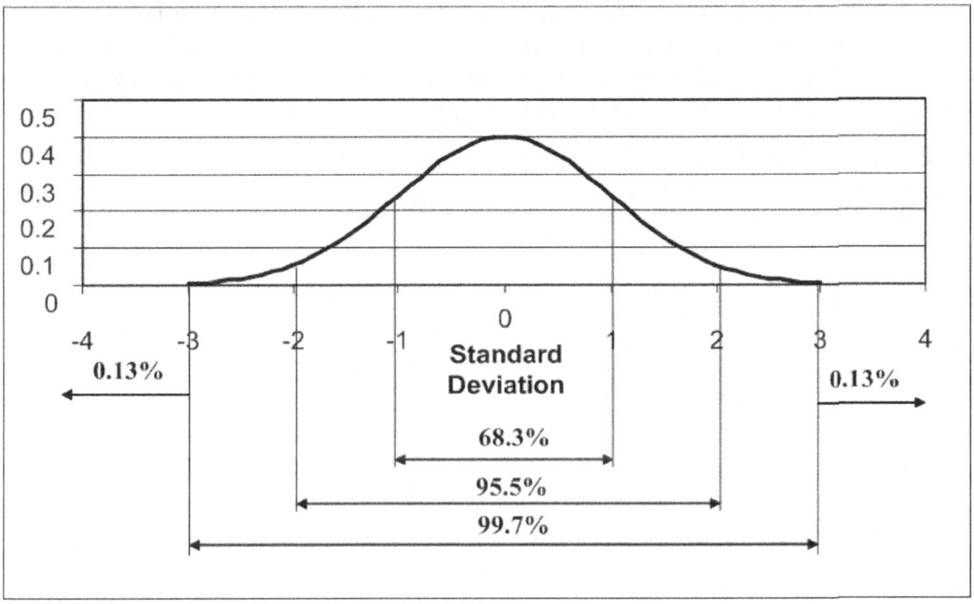

As shown on the graphic above,

about **68%** of values occur between **-1σ and + 1σ** about the mean,
about **95 %** of values occur within **2σ** of the mean and
about **99.7 %** of values occur within **3σ** of the mean.

Example: 40 students are taking an algebra course. On the mid-term test the mean score is 84 and the standard deviation is 7. About how many students failed the test if a failing grade is 70.

$70 = 84 - 2\sigma$

From the graphic above, about 95 % of students will fall within 2σ of the mean, 2.5 % will fall above $+2\sigma$ and 2.5 % below -2σ. Hence, about 2.5% of the students should have scores below 70. 2.5% of 40 is about 1 student.

Is it a normal distribution?

The techniques described in this section apply to populations accurately described by the normal distribution curve. Many phenomena in nature are normally distributed, but not all. How does one determine whether a population is normally distributed? You could do it by brute force; collect a lot of data and show that it is well distributed. But usually you must judge based on what you know about the phenomena. Is it likely that error will be evenly distributed, or are there some phenomena which will skew the data. Examples of factors that yield non-normal distributions are:

> If there is a lower bound or an upper bound close to the mean (many sample values are at or near the boundary), then it is likely not normally distributed.
>
> If a discrete event can cause two or more natural values, then the population is probably not normally distributed. Example: time to drive a distance that includes one stop light. Two peaks will exist, one for trips when the light is red and one for trips when the light is green.

The normal distribution can be a useful tool for improving judgment even in cases where we know, or suspect, the data is not normally distributed.

4.7 EXERCISE – NORMAL DISTIBUTIONS

You have conducted 10 experiments to measure the tensile strength of a new plastic part. The results are:

23.0, 24.3, 27.3, 25.2, 21.2, 22.2, 29.0, 27.3, 24.1, 27.0

a) Plot these points on the line below and by inspection, estimate the mean and the standard deviation
b) Estimate the fraction of parts which will have strength measurement less than 20.

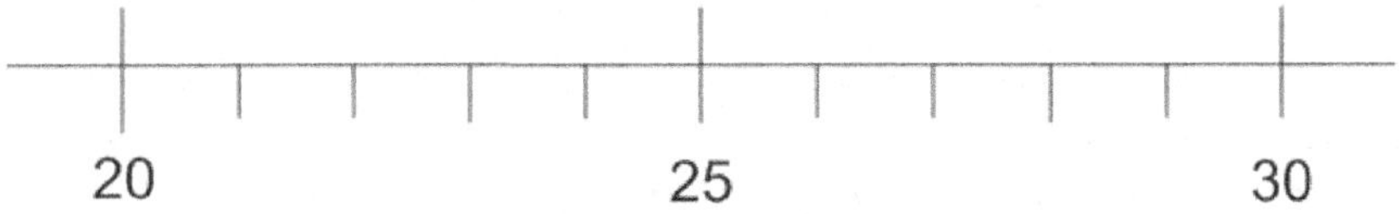

4.8 NORMAL DISTRIBUTION EXERCISE RESULTS

Mean

By inspection, the average or mean value appears to be about 25.

Standard Deviation

To estimate the standard deviation we have to observe the spread of the data. We expect about 70% of the data points to fall in the area between –1 std dev and +1 std dev about the mean. This appears to occur between 23 and 27 yielding an estimate of the standard deviation of 2. Further, we expect about 96 % of data points to fall between – 2 std dev and + 2 std dev of the mean. If the standard deviation is 2, then 96% of sample points should be between 21 and 29. On the plot below, that appears reasonable.

Fraction Below 20

21 is 2 std deviations below the mean. We expect about 2% of sample points to fall below that. 20 is about 2.5 standard deviations below the mean. We should expect about 1 % of measurements to fall below that value.

Relationship of Measurements to the Truth

When we say that we expect 1% of the measurements to fall below 20, that doesn't mean that 1% of the parts will have a tensile strength less than 20. The distribution of readings is partly due to a distribution of sample tensile strengths and partly due to a distribution of measurement error. If all samples have a tensile strength exactly equal to 25, we would still get the distribution described above if the measurement error has a standard deviation of 2.

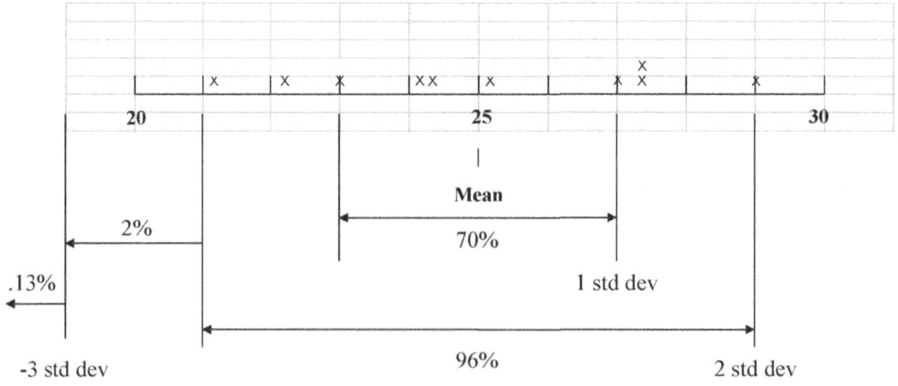

Checking the Results via Excel™

See Section 4.10 for a discussion of statistical functions in Excel™. Checking our results via Excel yields:

Average = 24.860
Standard Deviation = 2.376
Normdist = fraction below 20 = 2.0 %

4.9 THINKING IN TERMS OF DISTIBUTIONS

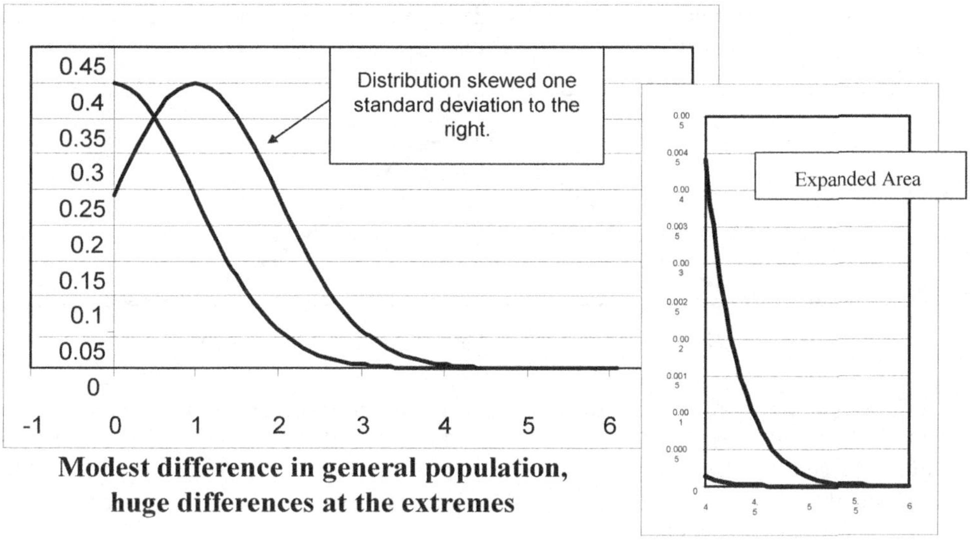

Modest difference in general population, huge differences at the extremes

The concept of the bell curve is very useful even if you never calculate such things. Much of the data we deal with every day is a sample from a distribution. Thinking of it in those terms yields insights.

Much of the data we are interested in occurs in the tail regions of the distributions (highly unlikely occurrence). Relatively small changes in the distribution as a whole can have large impacts in the tail areas. In the distributions above, one distribution has been skewed by one standard deviation. This has moderate impact on the population as a whole, but large implications at the extremes. In one case, very few samples had values greater than four. In the skewed case the number of samples with values greater than four has increased by orders of magnitude.

Measuring Safety in Industrial Plants

Want to know how safe a plant is? Don't look for records of serious accidents, they happen very infrequently even in poorly run facilities. Look for the frequency of minor events such as small spills, falls, trips, dropped objects, first aid incidents, etc. If the entire distribution is shifted to the right (more accidents of all types), the probability of a really serious accident is much higher, even if still small.

Why do Most Projects Underperform

Despite very large appraisal efforts, oil development projects typical do not produce as much oil as expected at project sanction. This is usually taken as evidence that project teams don't effectively account for uncertainty and make poor decisions as a result. It seems obvious that the average project result should be roughly equal to the sanctioned recoverables with some variation about the mean.

But that would only be true if the projects sanctioned were randomly selected from all available prospects. What actually happens is that projects with inflated recoverables estimates are more likely to be sanctioned.

Consider this simplified example. Suppose that there is a prospect that has actual recoverables of 100 MM (million) barrels of oil. Suppose further that many teams evaluate the prospect and that the standard deviation of the evaluation error is 20 MM bbls. The resulting distribution looks like this:

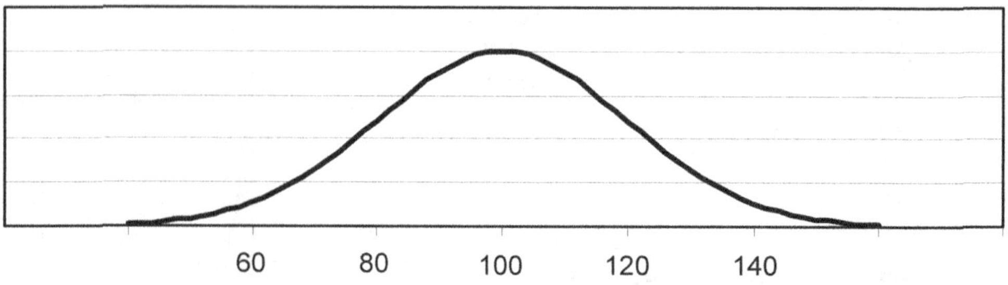

The average appraisal is 100 MM bbls recoverable. Half of the estimates come in under 100; 15% come in under 80 (1 std dev); 2% under 60 (2 std devs). If a team that estimates recoverables to be 60 MM bbls gets to execute the project, their project will dramatically exceed expectations. But their project probably doesn't seem economically viable and won't get sanctioned.

On the other hand, half of the teams come in over 100; 15% over 120; 2% over 140. Those projects are much more likely to be built, and they will obviously all underperform.

This built in bias for selecting projects with inflated estimates of recoverable oil seems to guarantee that oil projects will generally under-perform.

Application – Airport Security

We occasionally see news articles about people innocently getting guns and knives past airport security. Is that a problem? About 2,000,000 people pass through airport security every day in the US; 60,000,000 per month. Suppose one out of every 10,000 passengers accidentally brings a gun or a knife in his/her carryon baggage. Then 6,000 people every month will innocently approach airport security armed. How effective does airport screening have to be? I would feel very comfortable if the error rate were 1 in 1,000. That would be plenty adequate to deter any terrorist from attempting to smuggle a weapon on board. But it would also suggest that about 6 times per month, someone would get on board with a gun or a knife.

The fact that an individual occasionally gets a gun or knife past security is not in itself a newsworthy event and does not suggest that the system is flawed or ineffective.

4.10 NORMAL DISTRIBUTION CALCULATIONS VIA EXCEL™

Microsoft Excel™ provides statistical functions useful for evaluating normal distributions. A few are listed below.

RAND()
Returns a random number between 0 and 1.

AVERAGE(range)
Returns the arithmetic mean of the range.

STDEV(range)
Returns the standard deviation of the range.

NORMDIST(X,mean,standard_dev,TRUE)
Returns the normal cumulative distribution for the specified mean and standard deviation.

Normdist:	fraction of results with values < X
X:	is the value for which you want the distribution.
Mean:	is the arithmetic mean of the distribution.
Standard_dev:	is the standard deviation of the distribution.

NORMINV(probability, mean, standard deviation)
Returns the inverse of the normal cumulative distribution for the specified probability, mean and standard deviation.

Example

A series of experiments have yielded the results printed in column A. Determine the fraction of values expected to be greater than 10.0.

	A	B	C	D	E
1	6				
2	8		AVERAGE (A1.A16)		7.688
3	11				
4	9		STDEV(A1.A16)		1.702
5	6				
6	7				
7	9		NORMDIST(10, E2, E4,TRUE)		0.913
8	10				
9	5		NORMINV(E7, E2, E4)		10.000
10	9				
11	7				
12	8				
13	9				
14	6				
15	7				
16	6				

The formulas applied in column E are shown in column C.

The average value of the results is 7.688. The standard deviation is 1.702

The normal cumulative distribution at the value 10 (NORMDIST) is 0.913 which equals the fraction of samples expected to be less than 10. Hence, 0.087 fraction of samples (8.7%) are expected to be greater than 10.

As a check, the inverse of the normal cumulative distribution for a probability of 0.913 is 10.0 as expected.

4.11 SENSITIVITY – TORNADO DIAGRAM

Uncertainty can often be decreased by collecting more data. Data usually costs money. Only data that is worth the cost of collection should be collected. Unfortunately, it is difficult to estimate the value of data prior to collecting it. The tornado diagram is a simple method to identify the variables most likely to significantly affect your decision.

The tornado diagram below is for a home-based shortbread cookie business. Assume you have retired and wish to make some extra income by cooking shortbread cookies (via an old family recipe) to sell in local stores. You would like to earn at least $20 per hour in this business. Since you've never done anything like this before, there is a large uncertainty in many variables of the business. You've identified the major variables and roughly estimated the minimum, average, and maximum value of each.

Variable	Min	Average	Max
Bake Rate, number of 3 dozen batches per hour	3.5	4	4.5
Baking supplies cost, $/dz	2.25	2.6	3.1
Number of stores	4	7	10
Sales per store, no. of packs (3 cookies per pack)	12	18	28
% waste (undercooked, overcooked, ugly, eaten)	2	5	10
Sales Price, $ / pack	1	1.5	2
Store Cut, %	10	20	30

To generate the tornado diagram:

1) Estimate the profit ($/hr) when all variables are at their average value ($23.7/hr)
2) Taking each variable one at a time, perturb it first to its minimum and calculate the profit, then to its maximum and calculate profit. Perturbing the sales price to its min value with all other variables at their average yields a profit of $5.5/hr, whereas perturbing sales price to its max value yields a profit of $41.9/hr)
3) Arrange the results graphically as shown below with the variable having the largest effect on top

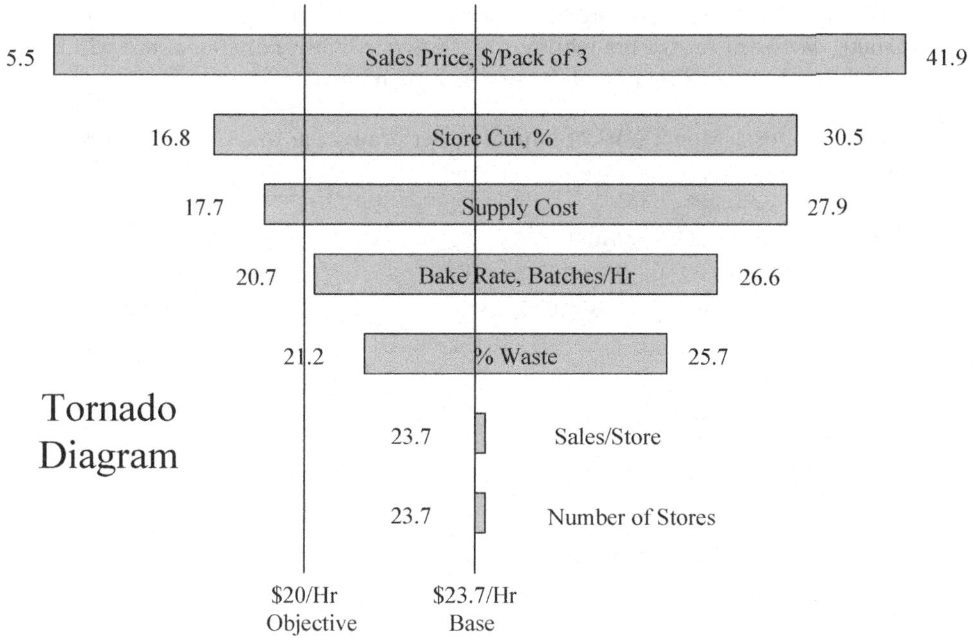

The tornado diagram is a useful tool for identifying the variables likely to have the biggest impact. It has some major limitations however.

Limitations of Tornado Diagram Method:

The tornado diagram method as described assumes the variables are independent. Every variable other than the one being perturbed is held at its average value. In many cases variables are not independent. In the shortbread cookie example, several variables are probably dependent. Sales/store is probably related to Sales Price (as the price goes up, sales rate will likely decrease). Waste may be related to Supply Cost; we will be more careful to avoid waste when the Supply Costs are high.

Even if variables are independent, it is unlikely that all but one variable will be at their average values while the one is at a minimum or maximum. All variables will vary about their mean in some fashion. The tornado diagram does not capture the impact of random variation of the variables. Monte Carlo Simulation does.

4.12 MONTE CARLO SIMULATION

Monte Carlo Simulation is a method of estimating the behavior systems with many variables. The method involves making multiple trials allowing each variable to vary according to its distribution.

Example – Two-Well Platform

Here is a very simple example. A project is to build a facility to produce two oil wells. Well 1 is predicted to produce 15000 BOPD (Barrels of Oil Per Day) with a standard deviation of 3000. Well 2 is predicted to produce 25000 BOPD with a standard deviation of 8000. We wish to size the facility for a flowrate that we are 90% sure will be greater than the actual production rate of the two wells combined.

Several lines of a Monte Carlo Simulation performed via Microsoft Excel™ are shown below.

		Well 1			Well 2		
	Mean	15,000	bopd		25000	bopd	
	Std Dev	3000	bopd		8000	bopd	
Trial	Random	Flow		Random	Flow		Sum Flow
1	0.490836127	14,931		0.347342	21,860		36,791
2	0.71293055	16,686		0.462015	24,237		40,923
3	0.902262313	18,884		0.660021	28,300		47,184
4	0.329504185	13,676		0.272758	20,164		33,840
5	0.271926458	13,179		0.962907	39,284		52,463
6	0.801066436	17,536		0.921704	36,333		53,869
7	0.484791493	14,886		0.266388	20,010		34,895
8	0.348575745	13,832		0.404979	23,076		36,909
186	0.009055853	7,940		0.53013	25,605		33,515
187	0.14886299	11,876		0.615009	27,339		39,215
188	0.194748625	12,418		0.217281	18,749		31,167
189	0.8375	[illegible]		[illegible]	[illegible]		33,926
190	0.0636	[illegible]		[illegible]	[illegible]		40,111
191	0.927482944	19,372		0.200081	18,269		37,641
192	0.021324188	8,919		0.205529	18,424		27,342
193	0.636179058	16,045		0.712715	29,491		45,535
194	0.125854713	11,561		0.783836	31,282		42,843
195	0.230463377	12,788		0.264881	19,973		32,761
196	0.592151425	15,699		0.605471	27,140		42,840
197	0.730471143	16,843		0.539785	25,799		42,642
198	0.168422022	12,119		0.444217	23,878		35,996
199	0.284009093	13,287		0.862393	33,729		47,016
200	0.464633627	14,734		0.775611	31,060		45,793
	Average	15,228			24,673		39,901
	St Dev	2919.648			7638.883		7983.939
	90% Flow	18969.64			34462.97		50133.13

@RANDOM()

@NORMINV(random,mean,std dev)

Average = sum of averages

Std Dev not = sum of std dev's

@NORMINV(.9,mean,std dev)

The simulation included 200 trials. For each trial two random numbers were generated. They were used to estimate the flowrate for each well via the NORMINV function. NORMINV returns a sample value (flowrate) based on a probability (random number) and the mean, and standard deviation of the distribution.

For each trial, the flowrates of the two wells are summed in the rightmost column. The average flow and standard deviation for the summed flowrate is calculated at the bottom of the page. A 90th percentile flowrate of 50578 BOPD is estimated.

Note that the average total flow equals the sum of the average flowrates of the two wells, but the standard deviation of the combined flow is less than the sum of the standard deviations. Hence, if we simply summed the 90th percentile flowrates of each well, we would oversize the facility.

Explanation of the spreadsheet:

Top of page:	The mean flow and standard deviation for each well
Column A:	Trial number. Total of 200 trials performed
Column B:	Random number for setting well 1 flowrate
Column C:	Flowrate of Well 1 for each trial determined via @NORMINV(random, mean flow well 1, std dev flow well 1)
Column E:	Random number for setting well 2 flowrate
Column F:	Flowrate of Well 2 for each trial determined via @NORMINV(random, mean flow well 2, std dev flow well 2)
Column H:	Sum of calculated well flows (Column C + Column F)
Below Trials:	Average and standard deviations of each well flow calculated as a check and 90th percentile flows. 90th percentile flows are estimated via NORMINV(.9, mean flow, std dev)

4.13 OPTIMIZATION

There exists a special class of decision problems in which the objectives and all the constraints can be expressed as mathematical functions. When this is the case we may be able to apply mathematical methods to determine the optimal solution. We'll briefly explore some of these techniques.

Calculus

For single differentiable equations, calculus can be applied to determine the minimum and/or maximum value of an equation. For example:

$$Y = 4 + 3X - 4X^2$$

The derivative of y with respect to x is 0 at a minimum, maximum or saddle points.

$dY/dX = 3 - 8X$

$0 = 3 - 8X$

$X = 3/8 = .375$

Y is maximized at X = .375

There aren't many opportunities to use such a simple approach. Most real life problems are described by systems of equations and more elaborate methods of solution must be employed.

General Form and Approach

For systems of equations the general form of the optimization problem is:

Objective statement such as:	Maximize Y
Objective Function:	$Y = F(x_1, x_2, \ldots.., x_n)$
Constraints:	Limits to possible values of x_i

Example: Maximize Y

$Y = 4 + 3x + 4z$

Such that: $x \leq 3$

$z = 2x$

Several methods are used to solve such problems depending on the characteristics of the equation set.

Linear Programming (LP):

If the objective function and all the constraint equations are linear, the method called Linear Programming (LP) can be used. LP is probably the most commonly used optimization method. It is easy to understand, relatively easy to program, and is very robust. The major drawback to LP is, of course, that all the equations describing the system must be linear. This occurs with some frequency in manufacturing systems.

Sequential Linear Programming:

LP can be used in fairly complex non-linear systems via a sequential linearization technique. Each equation is linearized (usually by finding the equation of the line tangent to the curve at a given point). The resulting LP problem is solved. The equation set is linearized again at the current solution point and the LP is solved again. This is continued until the method converges. Note that convergence is not guaranteed and, if you do get a solution it may be a local optimum rather than the global optimum.

Integer Programming:

Applicable if all the equations are linear and one or more variables take only integer values. This method is useful if some of the equipment is either on or off for instance.

Quadratic Programming

If the objective function is a quadratic equation and all the constraints are linear, a method called Quadratic Programming (QP) can be used.

Sequential Quadratic Programming

If the objective function and some or all of the constraint equations are quadratic, we can use Sequential Quadratic Programming (SQP).

Manual Search Methods

All of the above methods are iterative search methods. If the number of variables to adjust during an optimization is relatively small, less sophisticated manual search methods can be used. Spreadsheets have made this approach popular.

4.14 LINEAR PROGRAMMING EXAMPLE - OIL WELL PRODUCTION

Two wells are produced onto a platform. Maximum production rates from each well are given in the table below. Our objective is to generate as much revenue as possible from these wells, subject to three constraints:

1) The maximum total oil production rate (equipment limit) is 18000 BOPD (barrels of oil per day).
2) The maximum total gas production rate (equipment limit) is 40000 MSCFD (thousand standard cubic feet per day).
3) And there is a minimum gas production requirement (contract limit) of 20000 MSCFD.

	Maximum Production	
	Well 1	**Well 2**
Oil, BOPD	10000	15000
Gas, MSCFD	5000	25000

Objective: Maximize revenue, R
Oil Sales Price = \$20/bbl
Gas Sales Price = \$2/MSCFD
$R = 20*Oil + 2*Gas$

Constraints:

Oil Processing Limit = 18000 bpd
$$Oil = Oil_1 + Oil_2 \leq 18000$$
Gas Processing Limit = 40000 MSCFD
$$Gas = Gas_1 + Gas_2 \leq 40000$$
Minimum Gas Production per Contract = 20000 MSCFD
$$Gas = Gas_1 + Gas_2 \geq 20000$$

Solution:

Since the two wells together are capable of producing more oil (25000) than the platform is capable of treating (18000), we know that we will have to choke back (limit production) from at least one of the wells. We need to discover the percentage flow from each well that maximizes revenue.

$Oil_1 = P_1 (10000)$, $Oil_2 = P_2 (15000)$
$Gas_1 = P_1 (5000)$, $Gas_2 = P_2 (25000)$

Where: P_1 = Fraction of max flow from Well 1
Where: P_2 = Fraction of max flow from Well 2

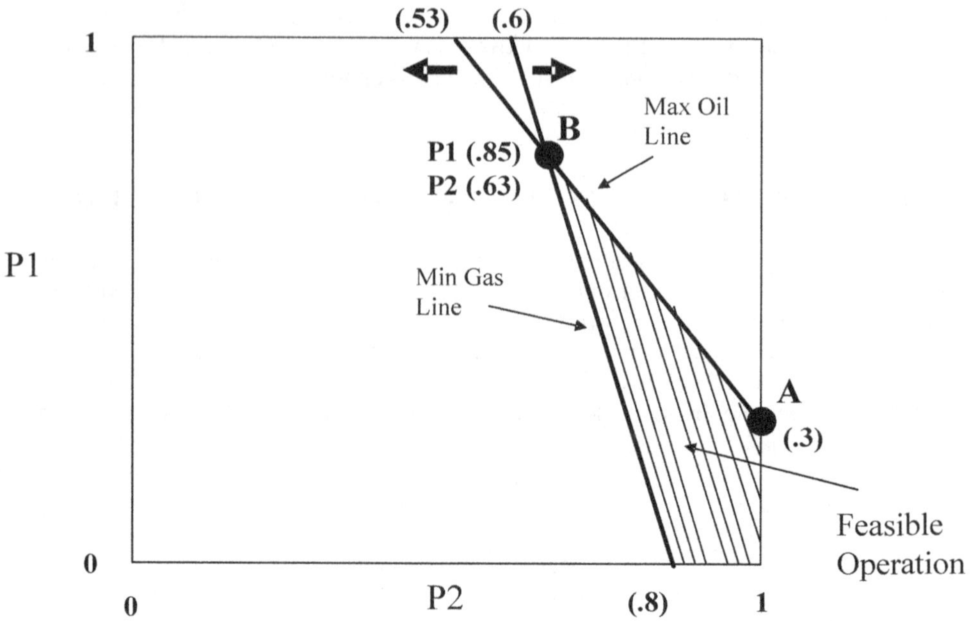

The plot above shows the entire problems space with P_1 on the Y axis and P_2 on the X axis. We know that we cannot run both wells at 100% open because the resulting oil production rate (25000 bpd) would exceed the facility capacity of 18000 bpd. We can plot in the Facility oil treating capacity limit:

$$Oil = Oil_1 + Oil_2 = P_1 (10000) + P_2 (15000) \leq 18000$$

Plotting two points of this line:

For $P_1 = 1$, $P_2 = 8000/15000$; $P_2 = .53$
For $P_2 = 1$, $P_1 = 3000/10000$; $P_2 = .30$

Hence, if Well 1 is wide open, Well 2 can only be run at 53% of capacity; if Well 2 is wide open, Well 1 can only be run at 30% of capacity. That line is plotted above (Max Oil Line) showing the areas of feasible and infeasible oil production. Every point on the line represents 18000 bpd production; to the right of the line is > 18000 bpd and hence infeasible.

The constraint on maximum gas production (40000) is of no concern because the two wells wide open can produce only 30000. The gas sales contract requiring a minimum of 20000 MSCFD of gas production does limit our alternatives. The line representing 20000 MSCFD gas production is plotted on the graph (Min Gas Line).

Operation must be to the left of max oil line and to the right of minimum gas line. We now need to find, within this feasible area, the well combination which yields the maximum revenue.

By observation it should be clear that that point will be on the upper or right boundary of the feasible space (on the Max Oil line). Only two points are contenders:

Point A: Well 2 @100% and Well 1 @ 30%

Yields Revenue of $ 413 M/day

Point B: Well 2 @63% and Well 1 @ 85%

Yields Revenue of $ 396 M/day

Point A is preferred.

We actually solved the problem graphically, but the LP algorithm works in a similar fashion. This is a relatively easy optimization problem that could be solved in numerous ways including manual search. But, even on a problem this simple, you might get the wrong answer if you don't use a structured technique. Imagine if you had 18 wells. Linear Programming solves this type of problem easily for problems with dozens or even hundreds of equations and variables.

4.15 LINEAR PROGRAMMING VIA EXCEL

Excel™ has an optimization solver which includes a linear programming algorithm. The graphic below shows the use of this solver to solve the well optimization problem on the previous page.

	A	B	C	D	E	F	G
1							
2			Max Oil	Max Gas	% Flow	Oil Prod	Gas Prod
3			bopd	mscfd		bopd	mscfd
4		**Well 1**	10000	5000	**.3**	3000	1500
5		**Well 2**	15000	25000	**1**	15000	25000
6		**Total**	25000	30000		18000	26500
7		**Limits (max)**				18000	40000
8		**Limits (min)**					20000
9		**Oil Sales Price**	20	$/bbl			
10		**Gas Sales Price**	2	$/mscf			
11		**Revenue/day**	413000	$/day			

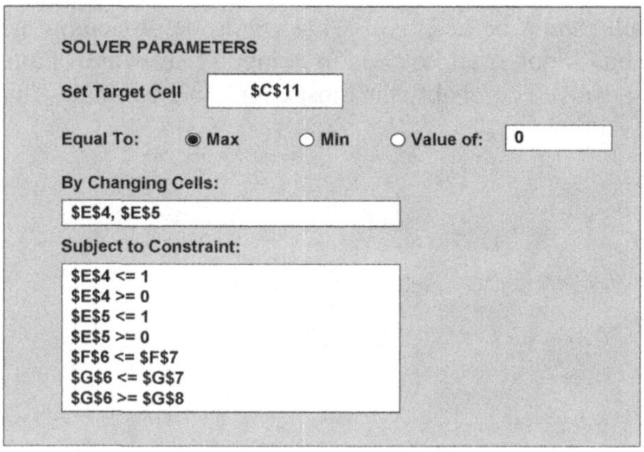

In rows 4 and 5 the spreadsheet calculates the oil and gas flowrates based on % of max flow from each well.

Flows are totaled in row 6.

Rows 7 and 8 identify the flow limits:

F7 contains the maximum oil flowrate (equipment limit)
G7 contains the maximum gas flowrate (equipment limit)
G8 contains the minimum gas flowrate (contractual obligation).

Cell C11 contains the objective function, revenue per day.

<u>Solver Block</u>

The inset solver block contains the LP model description. Starting from the top:

The target cell, C11 is our objective function.

In the second row "max" is checked indicating that we want solver to find the maximum value of the objective function.

Next we tell it to solve the problem by adjusting the values in cells E4 and E5 (% flow from each well).

The final scroll section contains the constraints. They are:

$E4 \leq 1$	Well 1 flow cannot be more than 100% of max
$E4 \geq 0$	Well 1 flow cannot be less than 0
$E5 \leq 1$	Well 2 flow cannot be more than 100% of max
$E5 \geq 0$	Well 2 flow cannot be less than 0
$F6 \leq F7$	Total oil flow must be less than the equipment limit
$G6 \leq G7$	Total gas flow must be less than the equipment limit
$G6 \geq G8$	Total gas flow must be greater than the contract min

The Most Common Linear Programming Error

Note the second and fourth constraint equations. They simply specify that the production from a well cannot be negative. That would be obvious to a human trying to solve the problem, but is not at all obvious to computer software. Failure to explicitly keep such variables positive is probably the most common error in developing LP models.

4.16 DECISION TREES

Decision trees are popular structures for evaluating alternatives. They apply the concept of expected value to make complex decisions.

DriveTek Decision
(from www.kellogg.nwu.edu/faculty/dana/htm/treeplan.htm)

You work for DriveTek, a computer storage device manufacturer. You have a contract that will pay DriveTek $200,000 for the successful development of a new tape drive. Company engineers have identified 3 methods of building a drive that will meet the customer's specifications: 1) mechanical drive, 2) magnetic drive, and 3) electronic drive. They are sure they can make the mechanical drive work (they've done it before), but it is expensive. The other drives would be cheaper to design, but there is risk that those efforts would fail. Best estimates of cost and probability of success are:

Option	Cost	Prob of Success
Mechanical	$120K	1.0
Electronic	$50K	.5
Magnetic	$80K	.7

If the option tried first doesn't work, other options must be implemented until a successful model is built. There is time to try all 3 if necessary.

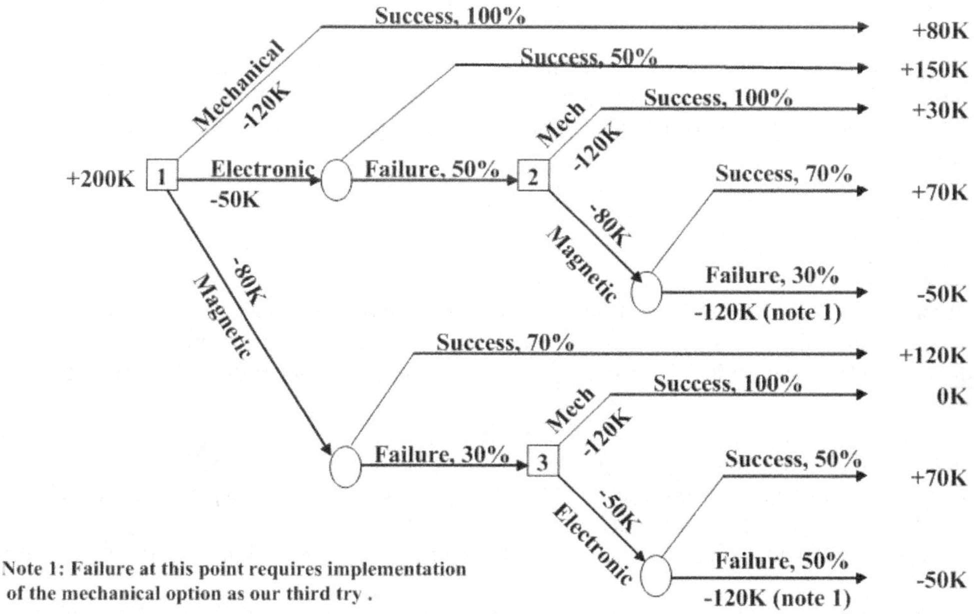

The decision tree describes the alternatives. The rectangles are decision nodes.

1	Decision node 1 represents the decision of which drive design to try first; mechanical, electronic or magnetic.

If the first choice is the mechanical option things are pretty easy. We know it will work. It will cost $120K. DriveTek will pocket an $80K profit.

The other options are more interesting. If the electronic option is selected, it may succeed (50% probability) in which case DriveTek pockets a nice profit of $150K. If the electronic option fails, there is another decision to be made at node 2.

The third leg emanating from node 1 represents the selection of the magnetic drive first. Attempting the magnetic drive will cost us 80K. If we are successful we'll make a 120K profit, if not we'll have to make another decision at node 3.

| 2 |

Decision node 2 occurs if the electronic option is tried first and fails. We have to decide which option to try next, mechanical or magnetic.

If the mechanical option is selected it will work. It will cost us another 120K and we will be left with a profit of 30K (200K – 50K for electronic try – 120K for mechanical try).

If we select the magnetic option at node 2 we will spend another 70K. There is a 70% chance we will succeed. If we do succeed we'll pocket a 70K profit. However, if the magnetic option fails we will have to implement the mechanical option as our third try. There is no decision node here because there are no other options to decide between. That will cost us another 120K and we'll wind up losing 50K on the project.

| 3 |

Decision node 3 occurs if we've tried the magnetic drive option first and failed. We now have to choose whether to try the electronic option second or go with the sure thing mechanical option second.

No surprises here. We can choose mechanical second, spend another 120K and break even. Or we can select the electronic option. Again we will be taking a risk. If the electronic drive is successful we net 70K; if not we implement the mechanical option and lose 50K.

Using the Tree

There is a great deal of information on this tree and it is presented in a logical format that apparently allows us to see the entire decision problem at one glance. For a problem with multiple decisions, it allows us to plan a strategy. Is it clear, from just looking at the decision tree above, what decision you would make first (node 1) or what strategy you would use?

Decision trees are solved from right to left using expected value to make each decision. This tree is solved in Appendix C. The solution tha maximizes the expected value is:

1. Choose the Electronic drive at decision node 1
2. If the Electronic drive design effort fails try the magnetic option at decision node 3
3. If the Magnetic Drive design fails design the Mechanical Drive

Weaknesses of Decision Trees

The strengths of the decision tree method are obvious. Complex decisions can be structured in a simple way. It allows for planning a strategy for solving problems involving multiple decisions. But the approach has major weaknesses as well.

Decision trees are useful for seeing the big picture of complex decision but they often overly simplify the decision problem. This tree is limited to evaluating options based on the expected value of each. Many other factors might be important including:

- Future prospects: Suppose magnetic drives are the wave of the future and this is an opportunity to gain experience on such drives. The potential loss might be well worth the risk in that case.

- Availability of resources: Presumably the people who would design the mechanical, electronic and magnetic drives work in different groups. Do all three groups have the resources required? Are the three groups equally committed to this project? Will you get the support you need from any and all groups? Suppose the magnetic group is eager to tackle this task and is actually lobbying for the job, whereas the electronics group views it as a distraction from more profitable opportunities?

The options mapped out on the Drivetek decision tree probably dramatically oversimplify the real life problem. Consider the electronic option. It is claimed that it will cost $50K with a 50% chance of success. Think about that claim for a moment; this cannot be the whole story. How realistic is it to think that we can spend $50K on an effort to develop an electronic drive and that at the end of that effort, and not before, we will be able to make an unambiguous assessment of success or failure. If the early effort is wholly unsuccessful we may be able to determine very early before spending much money at all that we will likely fail. On the other hand if we've spent the 50K and have a prototype that is very close to meeting all the specs the best option at that point might be to spend another $20K to improve the existing design rather than starting over on a magnetic option.

As the problem is stated in the decision tree the level of success apparently doesn't matter. Does success mean the same thing for all three? Presumably there are multiple criteria to be met (storage volume, speed, manufacturing cost, reliability, etc.). Different options will have different attributes. It could be that a successful electronic drive will be much more attractive to the client than a successful magnetic drive because it would be much more durable, for instance.

The problem is stated in the particular terms of cost and probability of success because those are the criteria that can be effectively managed via a decision tree. In this case it can be argued that the decision tree is used as much to define the problem as to solve it. And because of the limitations of the decision tree approach it appears that the problem, as defined, is not a meaningful problem to solve in this case.

For a more complete discussion of decision trees see Appendix C.

4.17 MODELING PREFERENCES

Expected Value

In making decisions under uncertainty, a common measure of value is the expected value. The concept is relatively simple; if I do this many times, what will my average win/loss be.

Expected Value, EV = P(1)*V(1) + P(2)*V(2) + ... + P(n)*V(n)

Where: P(i) = probability of i occurring
V(i) = value of i

Example: You are a finalist in a contest. You and 99 other people have received keys. One of the keys will start a new car worth $25,000 and that person keeps the car. Suppose you want to sell the key; how much is it worth?

EV = payoff * probability of winning
= 1/100 * $25,000
= $250

Expected Value Violations

There are some limits to the use of expected values in decision making. In the example above, we argue that the car key is worth $250. But you may keep the key even if someone offers you $500 for it. Another person may sell his key for $100. Whether you are willing to sell your key for less or more than the expected value says something about your risk tolerance.

How do we factor our tolerance for risk into our decision-making process? In the next few sections we will discuss three theories that cast light on the subject.

Utility theory provides a mathematical description of rational behavior. This includes a description of our risk tolerance, which can then be used to describe rationally our preferences in risky situations. For a short introduction to utility theory see Appendix D.

Prospect theory argues that utility theory does not accurately describe the way people actually behave. Since preference theory describes rational behavior, prospect theory suggests that we frequently behave irrationally.

Regret theory argues that we seek to avoid regret and disappointment. Avoidance of regret has value. By factoring in this value, regret theory provides a rational explanation for the apparently irrationality behavior described in prospect theory.

4.18 VALUE CURVE

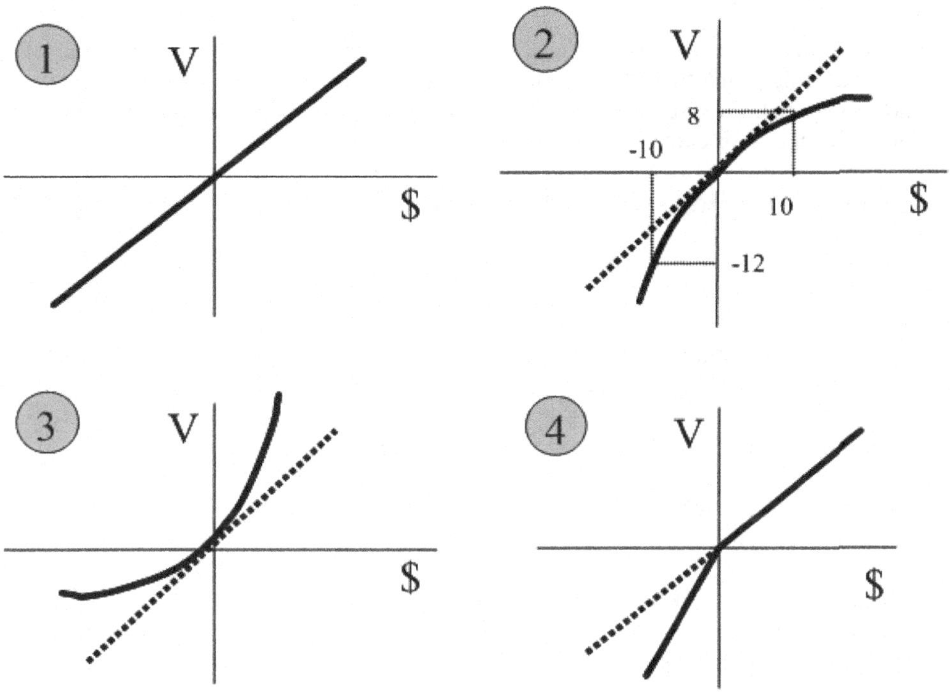

Risk tolerance can be expressed via value curves. The curves above represent different risk tolerance/avoidance behaviors. The curves feature dollars ($) on the horizontal axis and value on the vertical axis.

1 – Rational and Risk Neutral

The first curve represents a person who is completely indifferent to money and risk (risk neutral). This person would be indifferent to the game:

> Heads you win $10
> Tails you lose $10

2 – Rational, Risk Averse

The second curve is probably a better representation of common behavior. This person is rational and risk averse and will not play gambles unless the odds are in his favor. This person will not play the game:

> Heads you win $10
> Tails you lose $10

Because winning (+$10) is only worth 8 to him and losing (-$10) is worth –12. Hence, the expected value of the gamble from his perspective is

$$EV = .5(8) + .5(-12) = -2$$

3 – Risk Seeking

Few people are continuously risk seeking, but most of us are risk seeking some of the time. You could argue that playing the lottery and gambling in casinos (odds in favor of the house) are risk seeking behavior. (But you could also model such behavior as money intentionally spent for entertainment value.)

A truly risk seeking person would relish the game:

> Heads you win $10
> Tails you lose $10

and would seek opportunities to play it.

4 – Reference Effect – Risk Neutral away from Reference Point

Curve 4 shows the reference effect we discussed in chapter 2. Losing is more painful than winning is joyful.

This behavior mimics risk aversion if you are comparing a loss to a gain. However, in a choice in which both possible outcomes are positive, or both are negative, the straight lines suggest risk neutral behavior.

Expected Behavior

Given all these possibilities, how do we expect people to actually behave?

We expect people to be risk averse most of the time and we expect people to value loses more highly than gains (reference effect).

Question: **If value perception changes at the reference point, would you expect risk tolerance to change at the reference point also?**

If so, in which direction?

4.19 PROSPECT THEORY – REFERENCE EFFECT FOR RISK TOLERANCE

Tversky and Kahneman (P1) reported evidence of a reference point effect on risk tolerance, the tendency to be less risk averse in the region of loses than in the region of gains.

If this effect is true we expect the curve below the reference point to be steeper <u>and</u> to have less curvature, as shown on the curve at left or, in the extreme, to have the opposite curvature (risk seeking) as shown on the curve at right.

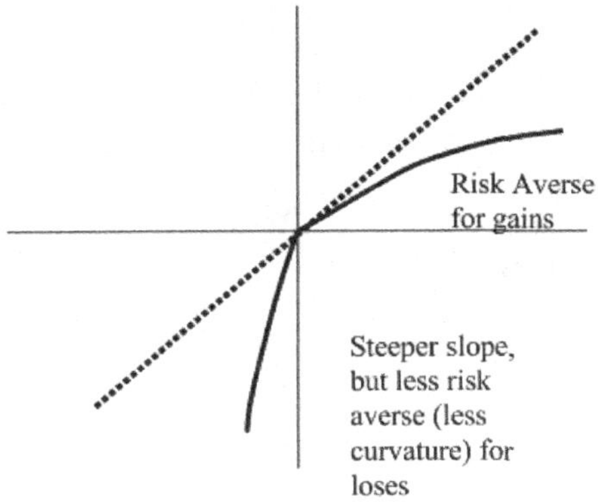

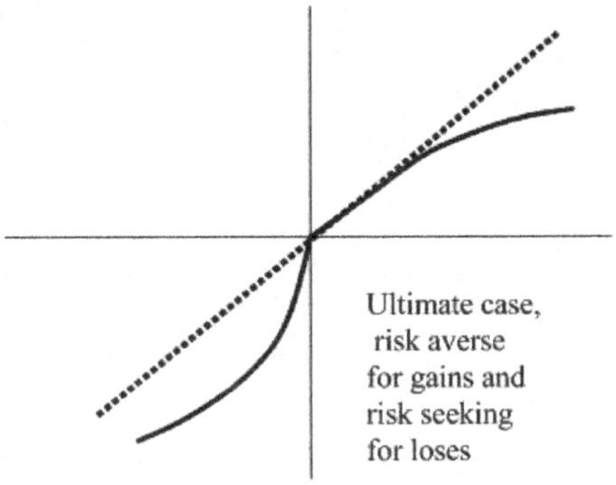

Tversky and Kahneman labeled this observed behavior prospect theory.

Example:

Consider these two scenarios:

1. You have the responsibility to perform a dangerous task (1/1000 odds of being killed). The task is not unpleasant, doesn't take very long, and is not illegal; the only negative is the danger.

   How much would you pay to avoid having to do it? ____________

2. One of your associates (not a close friend) has the responsibility to perform a dangerous task (1/1000 odds of being killed). The task is not unpleasant, doesn't take very long, and is not illegal; the only negative is the danger. He is willing to pay you to take the task off his hands.

   How much would he have to pay you to get you to do it? ____________

Did you give widely different answers to these two? Most people do. Yet, they are in essence the same problem. In both cases you are asked to determine the value, in dollars, of a 1/1000 chance of losing your life.

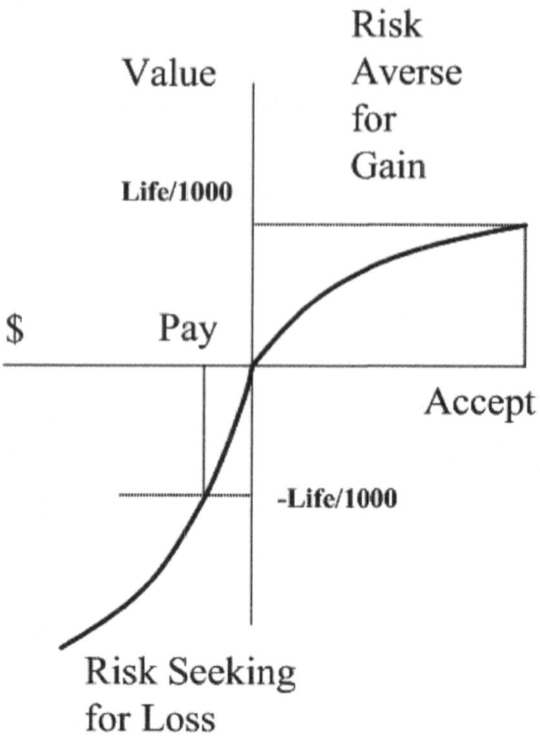

This example shows the reference effect on value quite clearly. It does not in itself provide evidence of the switch from risk averse to risk seeking at the reference point, but it does suggest such behavior. On the negative side, most people are willing to accept the risk rather than pay a large sum of money to avoid it. On the positive side, most people will not accept the risk unless they are very handsomely compensated. The interpretation of this is that we are risk averse over opportunities for gain, but are risk seeking to avoid loss.

Stock Market Application

Prospect theory can be used to model the behavior of stock owners. Suppose you've purchased a stock for $40/share and now it is selling for $30. Your reference point is 40 so you are now at –10 ($10 loss). You are expecting significant news from the company very soon (such as regulatory approval or disapproval for a major project). You think good news would bump the price up $10 and that bad news would knock it down another $10. Prospect theory predicts that you will hold on to this stock because the prospect of recovering the 10 that you lost is more valuable to you than the prospect of losing another 10.

Conversely, if you bought the stock at 20 and it is now selling at 30 you are likely to sell. It is unlikely that you would risk your $10 gain because the gain of another $10 is not as valuable to you as the potential loss.

Why does this Behavior Occur

As far as I know, Tversky and Kahneman didn't effectively address the reasons for this behavior. Several causes can be hypothesized; one is the power of commitment and responsibility in our society. If you agree to do something, or are responsible for doing something, that responsibility is a strong motivating force (has value).

Another explanation is the impact of regret and disappointment, which we will cover next.

4.20 REGRET

Regret theory explains the important behaviors described by prospect theory by hypothesizing that avoidance of regret or disappointment has value. If we include the value of avoiding regret in our value function, apparently irrational behavior appears more rational.

Before diving off into regret theory, we'll first discuss the phenomena of regret itself. The content in this section is taken largely from Gilovich and Medvec (P3).

Regret

Regret occurs due either to:

Sins of Commission:	We did something we now wish we had not done.
Sins of Omission:	We failed to do something we now wish we had done.

Regret occurs when the results of our action/inaction are worse than they could have been, or worse than we imagine they would have been. The fate of someone who dies in a plane crash after switching flights at the last minute seems more regrettable because we can easily imagine a better outcome.

The level of regret we feel is a function of how much worse the outcome was than it could have been or should have been.

Short-term vs. Long-term Regret

- People experience more regret over negative results from actions taken than from equally negative outcomes resulting from actions not taken.
- The biggest regrets in people's lives are usually things they failed to do rather than things they did.

These two claims appear to be in conflict. The fact that they are both true reveals something important about the experience of regret and our response to it. Action tends to create more regret in the short-term and inaction is more troubling in the long-term.

Why Regrets Shift with Time

- We tend to quickly take steps to correct the pain of regrettable actions (apologies, etc.). We do less to ameliorate the pain of inaction because it is not as strongly felt.
- We tend to find silver linings in regrettable actions; "That was a stupid thing to do, but I learned so much." We rarely learn anything from inaction.
- Our inaction is often caused by our fears or self-doubts at the moment of decision. These concerns fade with the passage of time and our failure to act seems more and more regrettable and harder to justify as time passes.

- The results of our actions are known and we have to deal with them. When we fail to act, usually we never learn how things would have turned out. Over time it is easy to imagine highly positive outcomes; "If only I'd opened that business I'd be rich and retired now."

Compelling and Restraining Forces

Forces that restrain action are less salient than the forces that compel action.

It is easy for individuals to assess why they acted.

It is much harder for individuals to understand why they failed to act.

When we try to induce another person to do something, we generally try to increase that person's motivation for doing so. Removing limiting factors is often more effective than adding inducements.

For example: Disney World is a very clean park. One of the reasons is the large number of trash cans and the fact that they are never full. Disney World makes it easy for their guests to throw trash in trash cans. That is far more effective than a promotional campaign encouraging the right behavior or a process for punishing bad behavior would ever be.

4.21 REGRET THEORY

This section is derived largely from the chapter by Bell in *Decision making* (4).

Suppose you own an unresolved lottery

(X,p,Y) Pays X with a probability p,
Pays Y with a probability 1-p, and
X is preferred to Y

The expected value of this lottery is $EV = pX + (1\text{-}p)Y$

The Value of Disappointment

If you are rational and risk neutral, you should be indifferent towards a trade of the EV for the lottery. If you keep the lottery and Y occurs you will be disappointed. Suppose your disappointment/regret is proportional to the difference between Y and Expected Value:

$$\begin{aligned} \text{Disappointment} &= d\,(EV - Y) \qquad (d = \text{constant}) \\ &= d\,\{\,[pX + (1\text{-}p)Y] - Y\,\} \\ &= d\,\{\,pX + Y - pY - Y\,\} \end{aligned}$$

$$\textbf{Disappointment} = \mathbf{d\,p\,(X - Y)}$$

This suggests that the greater the difference between X and Y, the greater our disappointment and the greater the odds of winning, p, the greater our disappointment if we lose. Both appear to be reasonable.

The Value of Elation

There is a flip side to this coin. If X occurs we feel elation. If elations is proportional to the difference between X and the Expected Value, then

$$\begin{aligned} \text{Elation} &= \text{constant}\,\{X - EV\} \\ &= e\,\{X - [\,pX - (1\text{-}p)Y]\} \end{aligned}$$

$$\textbf{Elation} = \mathbf{e\,(1\text{-}p)\,(X\text{-}Y)}$$

Elation is greatest when the difference between X and Y is greatest and when the probably of losing (1-p) is greatest.

4.22 MODELING PROSPECT THEORY OBSERVATIONS VIA REGRET THEORY

The S-curve behavior described in prospect theory can be adequately explained via regret theory. To show this I hypothesize:

1) We are typically somewhat risk averse over both gains and losses
2) We feel significant regret/disappointment over loses
3) We feel elation over gains, but elation is usually small compared to regret. This occurs partly because of the reference effect and partly because we usually expect to succeed (overconfidence bias). Failure is a surprise.

Generating a value curve from this model of behavior:

Hypothesis 1 yields a simple risk-averse curve as a starting point

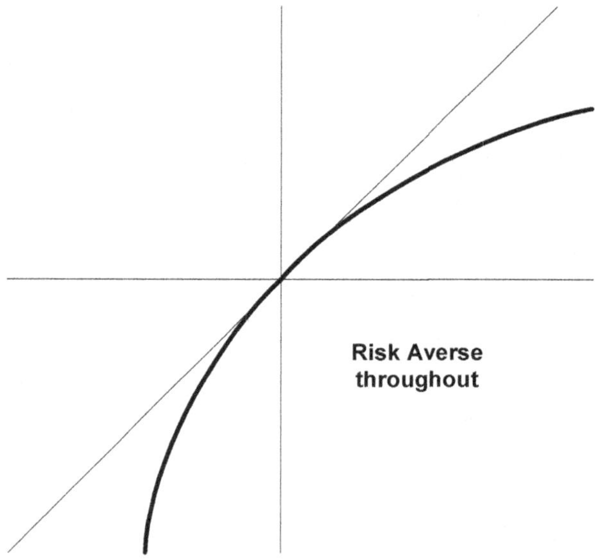

Including the values of regret avoidance and elation seeking yields a curve similar to the S-shaped curve from prospect theory.

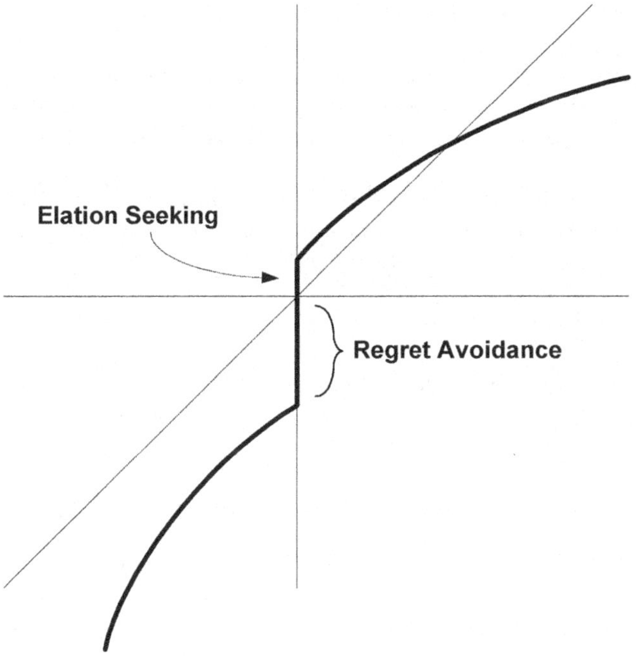

Whereas the S-shaped curve in prospect theory is entirely empirically derived, this curve has a reasonable theoretical basis. By predicting when people are likely to feel regret and how intensely that regret will be felt, it should be possible to predict behaviors more accurately with regret theory than with prospect theory.

4.23 SUMMARY – PREFERENCE MODELING

I've never seen preference modeling formally used to solve a real decision problem. I never use it myself. But I do keep the principles in mind. I think that is the proper use of this information for most decision makers.

Expected Value is often not the True Value

Expected value can be a strictly valid measure of value only if a decision is to be repeated many times. For one-time events expected value may be a reasonable measure for making decisions when the amount of money involved is relatively small, but not trivial, and you are risk neutral towards the decision.

Losses trump Gains – Understand your Reference Point

Probably the most useful insight from preference modeling is the simple observation that we overvalue losses and undervalue gains. This simple phenomenon can lead to wildly irrational behavior.

Use this observation to avoid unreasonable loss aversion and to help you reestablish your reference point when conditions change. If you have spent or lost money which is now irretrievable (sunk costs); that money should not figure into your current reference point.

Value Curves Change with Time and Circumstance

People (and businesses) do not live by a single value curve. We are risk averse at times and risk seeking at others. Circumstances change how we view risk and value money. Money appears to have less value when we are on vacation, for instance.

Retail sales people know that the first sale is the most difficult. Once they've got you committed to buying a suit, selling a shirt or a pair of shoes becomes easier. In effect, your value curve flattens out with the first purchase.

Risk Seeking to Avoid Loss

You are likely to take significantly higher risks to avoid loss. This can be very dangerous behavior. You may make a risky decision in an effort to recover from a previous poor decision and sink further and further into quicksand.

This behavior is very common on the golf course. A golfer who hits a poor tee shot into the trees may try a very risky shot between two trees to recover the lost shot. It is very unlikely that the golfer would try such a risky shot from the tee box.

Avoidance of Disappointment has Value

Probably the largest limitation of value curves is that they usually do not explicitly account for our desire to avoid disappointment. Avoiding disappointment has value to us and affects our behavior.

4.24 DECISION STRATEGY

It seems fairly obvious that the more important a decision is,
the more data we should collect,
the more attributes we should consider,
the more we should try to identify additional alternatives.

One of the most important decision most of us make is selection of a spouse. For such an important decision, it seems natural that we would collect large amounts of data about a great many possible mates, weight the data, build a spreadsheet to calculate the subjective utility (value) of each possible mate, and finally chose the one with the highest expected utility.

Did you use a spreadsheet approach to pick your mate?

Did your spouse use a spreadsheet to pick you?

Why not?

What's More Important, Deciding or Acting

Even if you can't put your finger on it, you know that there is a good reason you didn't use a spreadsheet to pick your spouse. It just isn't the right way to make that decision.

There are decisions in which selecting the best alternative is the most important factor in success and other decisions in which effectively implementing the selected alternative is more important than which alternative is chosen.

When selecting a spouse, love outweighs other considerations perhaps because the way you "live" your marriage is more important than the person you choose. If you are in love you are more likely to "live" the marriage in a way that yields success.

Decision Method Should Fit the Problem / Environment

It's clear that not all decision problems can be tackled via the same decision strategy. Selecting a project to invest in at work may require a decision strategy based largely on cold, hard facts. When you choose a new car you consider a combination of facts and emotional factors. When you choose the your friends emotional factors are predominant.

The decision strategy should fit the environment.

4.25 CLASSES OF DECISION STRATEGY

Gigerenzer (14) identifies four classes of decision strategy.

Unbounded Rationality

Unbounded rationality is the gold standard of decision processes. It is conceptually quite simple. Collect all relevant data and evaluate the data to determine which alternative provides maximum value. This is a laudable goal, but it requires unlimited access to information and perfect knowledge in how to evaluate the data. In the real world data is expensive and often ambiguous. And we cannot know all the implications of each alternative. Unbounded rationality cannot be applied by mortals. Unbounded rationality is irrational.

Optimization under Constraints (Bounded Rationality)

Recognition of the cost of data suggests a different strategy: collect data until the cost of collecting more data is greater than the gain that the additional data will provide.

That sounds good, but is not very actionable. You cannot know the value of data you haven't collected. At best you can estimate the potential value of additional data. But evaluating the probable value of all pieces of data not yet collected may be as hard a task as collecting the data. The problem is still unbounded. Bounded rationality is irrational.

Satisfycing

In real world problems the optimum solution is unknown and unknowable. Herbert Simon suggested satisfycing as a model of real world decision making. Here we seek enough data to identify one satisfactory alternative and stop the search at that point. Satisfycing provides an identifiable stopping point and provides for the possibility of limited (bounded search). Satisfycing can still be unbounded; if we set our aspiration level too high there may not be a satisfactory alternative.

Fast and Frugal Heuristics

Though satisfycing can simplify the search it can still be very time consuming and there is no guarantee that there exists a satisfactory solution or that the search will be worth the effort. The final search strategy to consider is fast and frugal heuristics. Heuristics are simply rules for stopping the search and making a decision. Via heuristics we make decisions without regard to whether the answer is either optimal or satisfactory. We simply follow the rules.

Examples of Fast and Frugal Heuristics:

Recognition Heuristic

> Given a choice between an alternative you recognize and one you don't, choose the one you recognize.

Take the Best Heuristic

> Choose the alternative that best satisfies the most important objective, ignoring all other objectives.

Fast and frugal heuristics greatly simplifying decision making, but at what cost in accuracy and effectiveness? How much effort should we ordinarily put into decision making?

It is reasonable to surmise that the more important the decision is, the closer we will approach bounded rationality. Conversely, for unimportant decisions we should feel very comfortable with simple heuristics. Let's investigate that thought.

4.26 PARADOX – LESS IS MORE

Our tendency is to use simple decision strategies for simple problems when little is at stake and to use more analytical strategies for complicated problems when a great deal is at stake.

Heart Attack Patient Risk Assessment

Gigerenzer (14) reports on the technique used by San Diego Medical Center (SDMC) for classifying heart attack patients as high-risk or low-risk. Clearly this is a complicated environment and very much is at stake. SDMC personnel measure and record up to 19 factors. Common sense dictates that all available data should be used to assess patient risk.

Indeed, programs have been written to assess risk based on statistical regression of all 19 factors, but the method used by doctors is illustrated below. It uses only three factors at most, and all three are yes-or-no questions. This may seem like a shortcut and an invitation to medical malpractice suits. But it is not. This simple method actually works better than statistical regression of all available data.

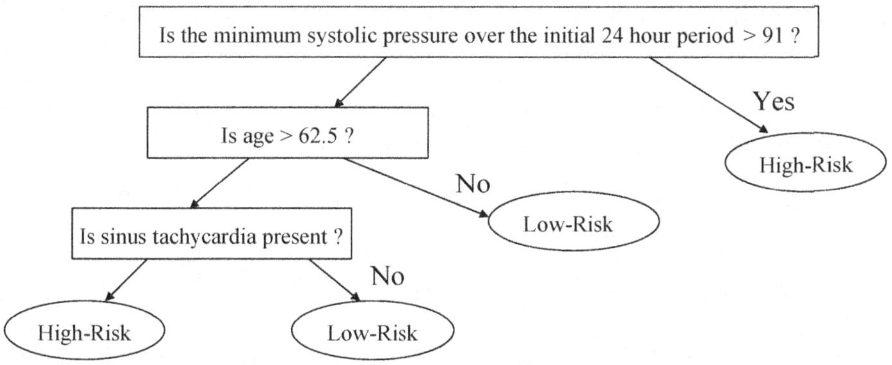

Using too much data actually degrades performance. The explanation for this is that regression yields, by happenstance, correlations that are true only of the sample population, not of the population as a whole. The more variables you use, the more likely it is that some spurious correlation will occur. Further, the more data you collect and use, the more likely it is that some of the data will be inaccurate.

Accurate collection and assessment of the most important data is more important than being comprehensive.

Fitting the Decision Strategy to the Environment

The risk assessment for heart attack patients is an important decision, but is made via a very simple method with little data. The decision strategy used by SDMC works well because it fits the environment.

Fitting the Decision Model to the Environment

Simple heuristics will outperform complicated models in simple environments.

Complicated decision strategies only make sense when

1. The environment is complicated,
2. Our model of the environment accurately reflects the complexity of the environment, and
3. We have enough data, and accurate enough data, to effectively use the model.

Most environments we deal with are actually fairly simple. Even in complex problems, a few variables often dominate. In those cases, accurate assessment of the important information is more important than comprehensive assessment of all possible data.

4.27 WORKPLACE DECISION STRATEGIES

The workplace decisions that we recall vividly are likely to be the ones in which we invested a great deal of effort, collected a great deal of data, and carefully evaluated the data to yield the best possible decision. Such efforts are rare, if only because we don't have the time to do that very often.

Most workplace decisions are made quickly with a minimum of data. At least that's the case when one person is responsible and accountable for the decision.

Selling a Decision

When you have to convince others, or "sell" your decision, you are likely to collect a great deal of data, massage the data, consult with experts, etc.

Note that this effort often occurs <u>after</u> you've made your decision. The data collection and evaluation is done to validate the decision rather than to help you make it. Even in cases where we don't have to defend a decision, we often make a decision before collecting data and then collect data to validate the decision to ourselves.

Collecting and evaluating data after we've made a decision seems backwards of course. It is interesting that this effort expended in defending the decision rarely causes us to change our minds.

Heuristics in Decision Making, Unbounded Rationality in Decision Selling

The way we make decisions often resembles fast and frugal heuristics. The way we defend decisions often resembles unbounded rationality and/or optimization under constraints.

<u>Defending Decisions made via Fast and Frugal Heuristics:</u>

You may have made the decision in a way that accomplishes one specific objective which is important for you or your team (Take the Best heuristic). If that objective is not of prime importance to other teams, you have to use other arguments to defend your decision.

Or you may have made the decision based on previous history; the way you've always done it (Recognition heuristic). If you have to defend it to people without the same history, you may need to make other arguments.

4.28 CHAPTER SUMMARY

This chapter addresses the selection of the favored alternative given a set of objectives.

This chapter contains short discussions of:

- Bayesian reasoning,
- Probability,
- Decision trees,
- Utility theory (the mathematical description of rational behavior)
- interesting psychological insights into seemingly irrational behavior (prospect theory and regret theory), and
- Discussion of decision strategies.

We observed that simple random behavior often looks like patterns. Hot streaks in sports are easily explained as random occurrences, for instance.

Bayesian reasoning make effective use of background probabilities in making sense of new data. Since failure to accurately account for background probabilities (base rates) is a typical bias, Bayesian reasoning can significantly improve our understanding of natural systems.

Expected value is a common decision criteria. The concept of expected value is simply "if I do this many times, what is my average gain or loss". The concept of expected value is frequently applied via decision trees.

But there are limits to the use of expected value in decision making. You may be willing, even eager, to buy lottery tickets even though they are not a good deal (risk seeking behavior). You may pass on a risky bet with a positive expected value if the potential loss is too great (risk averse). Several preference theory models attempt to explain violations of expected value. The best known preference theory model is prospect theory. Prospect theory argues that we are typically risk averse over potential gains, but often risk seeking to avoid potential losses. Regret theory explains the same behavior patterns by arguing that avoidance of regret has value to us.

The chapter concludes with are short discussion of four possible decision strategies:

- Unbounded rationality
- Bounded rationality (optimizing under constraints)
- Satisfycing
- Fast and Frugal Heuristics

Unbounded rationality requires collection and evaluation all possible data about a problem. Since that is impossible, the method is impossible and irrational. In bounded rationality we collect data until the cost of new data exceeds its worth. Since there is no practical way of determining the value of uncollected data, bounded rationality is also irrational. Satisfycing ends search when a satisfactory alternative is found. It is a pretty effective description of much engineering decision making. Satisfycing may be unbounded if no satisfactory alternative exists, or none can be found within the time and resources available. Heuristics stop the search when a predetermined stopping point is achieved. These techniques can be much faster. In some environments they will perform as well as other, more rigorous methods.

4.29 ANSWERING THE FOUR QUESTIONS

What observations from this chapter can we make about our four key engineering questions?

Why do so many changes occur late in projects?

Fast and Frugal Decision Making:

Decision made easily and quickly via fast and frugal techniques may be found wanting if evaluated later via analytical techniques.

Different Stakeholders have Different Levels of Risk Tolerance

Decisions involving risk are difficult. Different people make different decisions on what level of risk to accept. Because these decisions are hard they drag on or are revisited and often cause changes late in the project.

Why do we have so many problems at interfaces?

Selling a decision across interfaces:

Decision making within a team is often relatively straightforward. Team members speak the same language and have the largely the same objectives. A decision easily reached within a team may be very difficult to "sell" to other teams. The strategy used for "selling" the decision is likely to be different than the strategy used to make the decision. Decisions made via fast and frugal techniques within the team may have to be sold to others using analytical techniques.

Why do we repeat past mistakes?

Frugal decision methods make learning difficult and unreliable:

When we apply a detailed analytical effort to make a hard decision, we have a basis for investigating what went wrong with the analysis. When we make decisions via fast and frugal heuristics there may be little information in the decision model to guide us in making a better decision the next time.

Why do we so often finish behind schedule?

Applying inappropriate decision strategies:

A complex problem with a great deal at stake and multiple stakeholders with different objectives is likely to demand a great deal of effort. A typical or simple problem with little at stake should be solved more easily and quickly. Different decision problems suggest different strategies. We waste money and time when we apply too much effort to a simple problem.

Multiple Teams and Multiple Contractors all need to finish on time

Complex projects depend on multiple design teams, vendors and contractors all finishing their parts on time. The more players there are, the more likely it is that at least one of them will be late.

Chapter 5
Acting

Sources:

The most important material in this chapter is about theories of action. That material is based on the action science research of Chris Argyris and others; see *Flawed Advice and the Management Trap* (1), *On Organizational Learning* (9), the chapter by Argyris in *Decision making* (4) and *The Reflective Practitioner* (32) by Donald Schon.

This chapter also contains information on situations that exert influence on us taken from Robert Cialdini's *Influence* (13).

The final sections of this chapter deals with the impact of emotions on our actions. That material is mainly from *Reason and Passion* (12) by Lazarus and Lazarus.

Preface:

We seek decisions that result in effective actions. To be effective the decisions must actually be implemented and the results must meet our expectations without any negative, unintended side-effects. But many apparently excellent decisions are never implemented, and many actions have consequences different than and/or in addition to, those intended. Some decisions are not implemented because they are not actionable (cannot be implemented). In this chapter we'll investigate the properties of actionable and effective decisions.

It's not who you are inside that counts;
it's your actions that define you.

Christopher Nolan, Batman Begins

The number one problem with learning to play golf is our natural tendency to do the obvious thing to solve a problem.

Percy Bloomer

CHAPTER 5 CONTENTS

5.1 INTRODUCTION
5.2 ACTIONABLE AND EFFECTIVE
5.3 EXERCISE: ACTIONABLE AND EFFECTIVE ADVICE
5.4 EXERCISE RESULTS: ACTIONABLE AND EFFECTIVE ADVICE
5.5 DEVELOPING ACTION PLANS
5.6 IDENTIFYING POTENTIAL ERRORS VIA USE CASES
5.7 IDENTIFYING POTENTIAL ERRORS IN IMPLEMENTING PROCEDURES
5.8 THE PLANNING FALLACY – SENSEMAKING IN PLANNING
5.9 HUMAN ERROR MODELING
5.10 FAILURE MODES AT THE SKILL-BASED LEVEL
5.11 FAILURE MODES AT THE RULE-BASED LEVEL
5.12 FAILURE MODES AT THE KNOWLEDGE-BASED LEVEL
5.13 SYSTEMS THEORY - UNINTENDED OUTCOMES
5.14 THEORY OF ACTION – Espoused Theory
5.15 THEORY OF ACTION – Theories in Use
5.16 MODEL I vs. MODEL II BEHAVIOR
5.17 RIGHT-HAND LEFT-HAND EXERCISE
5.18 EVALUATING THE RIGHT HAND-LEFT HAND EXERCISE
5.19 EMIGRES ENJOY FREEDOM TO PRAY – AND TO WORRY
5.20 EXERCISE RESULTS
5.21 EXERCISE – REFLECTING ON GROUP PERFORMANCE
5.22 EXERCISE QUESTIONS– GROUP PERFORMANCE
5.23 EXERCISE RESULTS – GROUP PERFORMANCE
5.24 GROUP PERFORMANCE EXERCISE – A DIFFERENT STORY
5.25 IMPACTS OF MODEL I BEHAVIOR
5.26 REFLECTIVE PRACTICE
5.27 RELATIONSHIP BREAKDOWN IN THE WORKPLACE
5.28 COGNITIVE BASIS FOR MODEL I BEHAVIOR
5.29 INFLUENCE
5.30 PROTECTING YOURSELF IN A CROWD
5.31 EMOTIONS IN DECISION MAKING
5.32 CAUSES OF EMOTIONS
5.33 METHODS OF COPING WITH EMOTIONS
5.34 REFERENCE EFFECT – POSITIVE AND NEGATIVE EMOTIONS
5.35 CHAPTER SUMMARY
5.36 EVALUATING BUSINESS ADVICE
5.37 ANSWERING THE FOUR QUESTIONS

5.1 INTRODUCTION

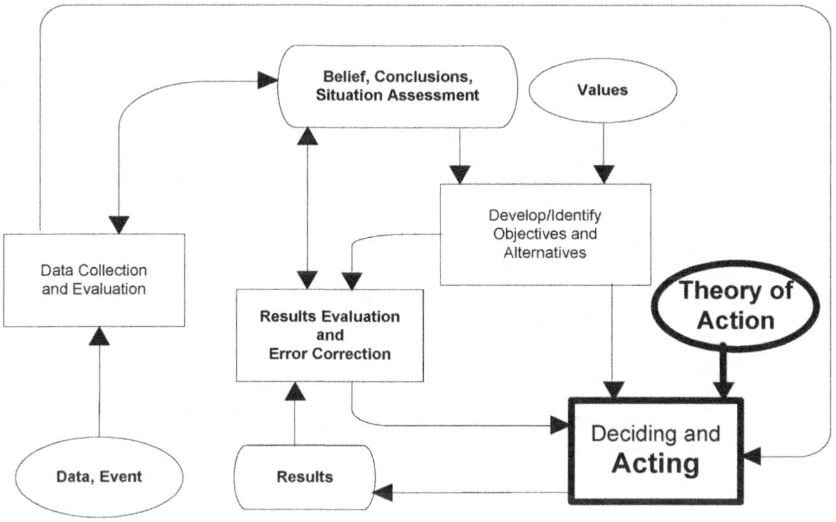

Most studies of decision making stop with the selection of the preferred alternative. The assumption apparently is that implementation is trivial once the alternative has been selected. But we care about the outcome, not simply about identifying the most promising alternative. And it can be difficult to implement even the best of decisions and our actions don't always achieve our intended results.

Three common problems face us in the implementation phase:

1) failure to act on a decision,
2) resistance by others when we try to act, and
3) objectives not achieved and/or undesirable side effects occur.

To achieve our objectives we should seek alternatives that are actionable and are likely to be effective:

Actionable: A typical member of the target audience is capable of implementing the proposed alternative.

Effective: If implemented as planned, the alternative will achieve the desired objectives without generating undesirable and unanticipated side-effects.

In group settings, there are social factors that limit our actions. And the reactions of others (also subject to social limits) will impact the effectiveness of our efforts. These social limits are partially explained via the concept of theories of action.

The effectiveness of actions has several components. One of the main contributors to ineffectiveness of actions is unintended consequences. We will take a systems view of actions in order to gain some understanding of such effects.

5.2 ACTIONABLE AND EFFECTIVE

Actionable Plans

There are some action items that we cannot accomplish. A plan may be unactionable for a number of reasons:

Skills, Knowledge:	You lack the physical skills and/or the knowledge to take the action.
Lack of authority:	You lack the authority to effect actions that must be done by others (the action may still be actionable if you can convince them to do it).
Interpersonal skills:	The required action may require communication skills, interpersonal skills or a level of self-confidence that you don't possess.
Ambiguity:	The action item is too ambiguous, or is not formulated as an action.
Social limits:	General cultural or social limits to behavior and specific limits created by our membership in a group such as a class, team, or church make it difficult to take some actions. In particular, defensive reactions of others often make it difficult to implement some actions.

The first three items on the list are fairly obvious limits. The others two are less so in concrete situations. Ambiguity is an especially prominent feature in business advice. We are frequently told for instance that it is important to "build a sense of trust". This is unquestionably an important goal in many situations, but it is not actionable advice as no action is suggested and it is not clear what actions would be effective.

Social limits are more interesting and more pervasive. Our culture places significant limits on our behavior. In a group setting your actions may be limited by your standing in the group and the opinions others in the group hold of you.

Effective Actions

Actions are effective if they achieve our desired objectives without generating undesirable and unanticipated side effects. Ineffectiveness typically results from one or more of the following three causes:

Errors	Some of our plans are simply wrong and do not achieve the desired results for reasons which are obvious. For instance, an architect designs a building for low density occupancy, unaware that it will be used for large company parties. As a result, the air conditioning system is undersized. The correction of such simple errors is straightforward (though not necessarily easy).

Action plans can get very complex; it is easy to get them wrong. We'll investigate the use of the means objectives network as a guide to developing complex action plans.

More interesting are cases when our plans are largely sound, yet errors are made in the implementation. The work of James Reason on human error (28) provides guidelines for identifying likely errors in the implementation of plans.

Systems effects

For an excellent treatment of systems theory read *The Fifth Discipline* by Peter Senge (20). When we plan actions we often don't consider the impact of those actions on the entire system. Often the system interactions undo all the good we intended. Systems reactions are especially troubling when, as is often the case, the good is achieved quickly and there is a time delay before the harmful side effects are felt.

Defensive reactions

In group settings, many actions are undone by these defensive reactions. Defensive reactions often seem unreasonable and irrational to us. Understanding defensive reactions and developing the ability to predict them is crucial to developing effective and actionable plans.

5.3 EXERCISE: ACTIONABLE AND EFFECTIVE ADVICE

In her book *Hardball for Women,* (24) Dr. Pat Heim suggests several methods for women to deal with conflict. One of her suggestions is **"use humor to defuse conflict when facing one-on-one conflict".**

Spend a few minutes answering the following questions:

1) At first glance, does this advice seem reasonable?

2) Is "use humor" actionable? Do you understand what it means? Can you do it? Can the average person do it?

3) If/when it is actionable, is "use humor" likely to be effective? Will it end the conflict? Will it end it in a productive way?

5.4 EXERCISE RESULTS: ACTIONABLE AND EFFECTIVE ADVICE

Does the advice seem reasonable?

At first glance, use of humor to defuse a conflict does seem to be reasonable advice?

Is the advice actionable?

The answer to this question might depend on what she means? Certainly we can all tell a joke. But that seems inappropriate. Animal sounds are funny, but probably would be frowned upon. We must assume she means using humor relevant to the situation. That's much more difficult. Robin Williams could probably defuse almost any conflict with his incredibly quick wit and effective delivery and timing. The rest of us would have less success thinking of something to say and then making it sound funny. We may, on occasion, have a very humorous thought that can bring levity to an otherwise tense situation. At first glance it seems reasonable to propose use of humor in that way. But, when it happens, it is usually spontaneous. It is a different thing to suggest that, during a conflict, we can deliberately come up with something funny to say that is appropriate and well timed.

Your position in the organization may impact your ability to be funny. The boss, the esteemed subject matter expert, or the team's natural leader might be able to joke in situations where subordinates should not.

Is the advice likely to be effective?

I remember as a child hearing my parents arguing. The argument had something to do with my oldest sister who has a fiery temper, as did my dad. At one point my mother exclaimed "well, it's clear that she's your child!" To which my dad answered, after a pregnant pause, "I thought they all were." My mother immediately burst out laughing and the argument was over.

Certainly, humor can be effective in defusing an argument. But is it likely to be?

We can easily be misled by previous, isolated experience. If it occurs spontaneously, is appropriate to the situation, in an argument that is not too intense, humor may work very well. But there is clearly a large downside.

The effectiveness of humor depends on the reaction of the other party. Your adversary may laugh and you may proceed to have a thoughtful and fruitful discussion, or may take offense and may assume that you don't care. Use of humor may harm the long term relationship.

If humor is used as a means of avoiding discussion of the real issues it may be effective in the short term (stops the argument), but ineffective in the longer term.

Note that use of humor requires deception. You will not say "we are having a conflict and I will inject some humor now to defuse it." If your opponent notes the humor and challenges it; "This is an important discussion and I don't think you should try to avoid or minimize it with humor", you would deny that intention. Deception, however well intentioned, is a dangerous communication tactic.

Comment - Why so much business advice is bad advice:

The advice given by Dr. Heim appears to be bad advice; the average person is incapable of applying it, and it is likely to be ineffective even when applied correctly. This is not unusual in the field of business advice. One of the reasons for this is that our standards for evaluating business advice are too lax.

Any new scientific theory must be supported by calculations and/or data from controlled experiments. In business advice the assumptions made are rarely tested and often are not testable. Much of the advice given is appealing and compelling; often it appears to be obviously true. Because of the appearance of obvious truth we don't insist on testing. It seems obvious that "use humor" is a good idea until we subject it to careful evaluation.

5.5 DEVELOPING ACTION PLANS

On the next page is an action plan for preparing for gas buyback – a subject we've discussed before (section 3.15). The plan is a means objectives network. We studied means objectives networks in Chapter 3 as a means of making decisions. I find them very useful as a means of planning the implementation of decisions.

Constructing the Plan

Start: Construction of the network is done from right to left starting with the overall fundamental objective or a set of lower level fundamental objectives.

Next: Identify all objectives or tasks necessary to achieve the overall fundamental objective.

Continue: For each task or objective added, continue to build the network by adding tasks or objectives necessary to achieve it.

End: Stop when you reach something that is already done, is already well planned, or is obvious and needs no other explanation or planning.

In this example the overall fundamental objective is "Safe and Effective Gas Buyback." Working our way to the left we find that eleven means objectives must achieved in order to achieve the overall fundamental objective. Some of the tasks need further definition or have multiple parts. For example, to achieve trained operators we need to have operating procedures available, develop a training program, deliver that training program, etc.

In this format it is easy to check the list for completeness; simply check that the inputs to each node are complete. And it is easy to track progress; the shaded blocks on the diagram indicate actions already accomplished at one point in time.

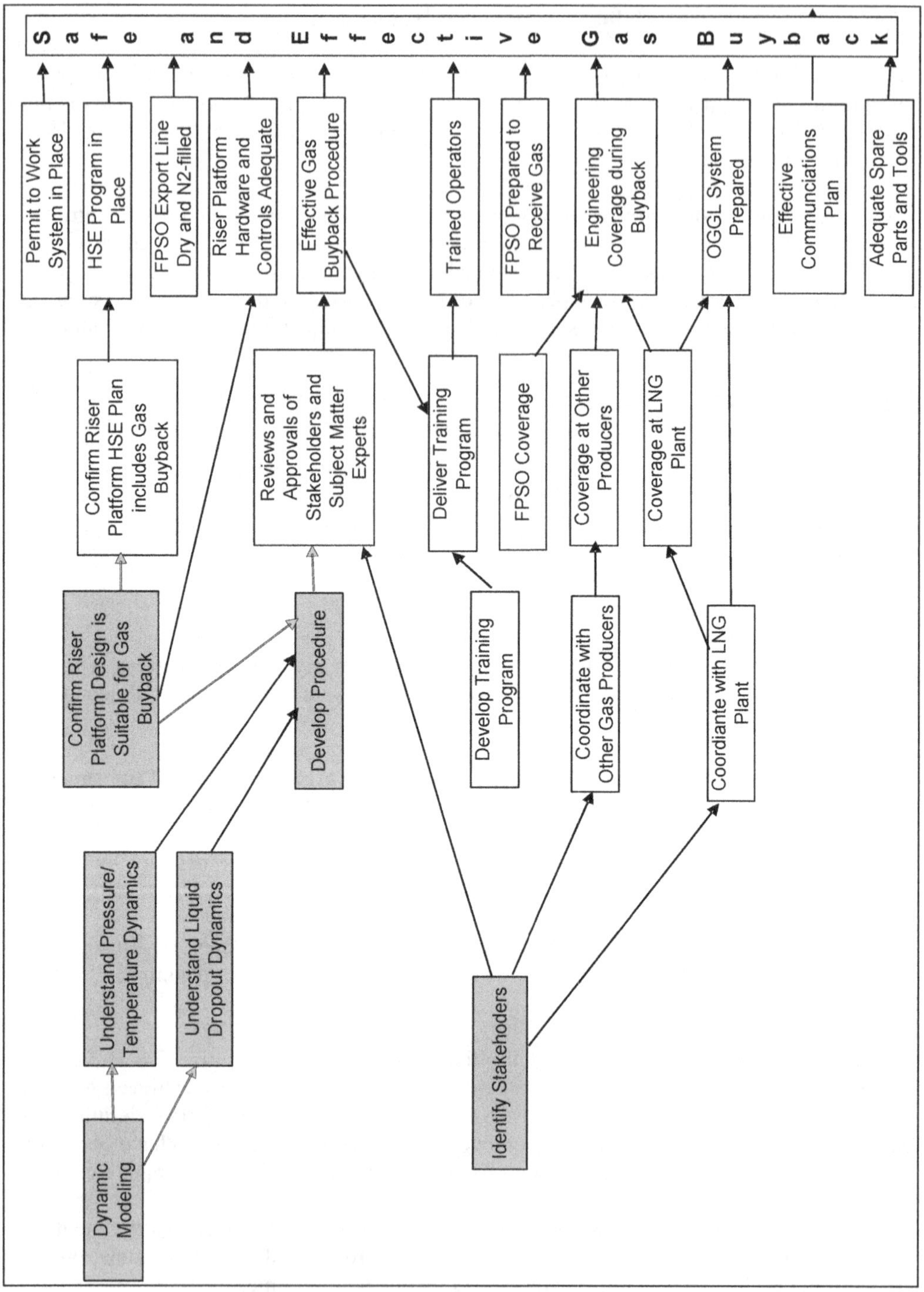
Safe and Effective Gas Buyback
Permit to Work System in Place
HSE Program in Place
FPSO Export Line Dry and N2-filled
Riser Platform Hardware and Controls Adequate
Effective Gas Buyback Procedure
Trained Operators
FPSO Prepared to Receive Gas
Engineering Coverage during Buyback
OGGL System Prepared
Effective Communciations Plan
Adequate Spare Parts and Tools
Confirm Riser Platform HSE Plan includes Gas Buyback
Reviews and Approvals of Stakeholders and Subject Matter Experts
Deliver Training Program
FPSO Coverage
Coverage at Other Producers
Coverage at LNG Plant
Confirm Riser Platform Design is Suitable for Gas Buyback
Develop Procedure
Develop Training Program
Coordinate with Other Gas Producers
Coordiante with LNG Plant
Understand Pressure/ Temperature Dynamics
Understand Liquid Dropout Dynamics
Identify Stakehoders
Dynamic Modeling

5.6 IDENTIFYING POTENTIAL ERRORS VIA USE CASES

The use case process is briefly described in section 3.6 where it is proposed as an aid in identifying objectives. Here we use the technique to identify potential error in the application of procedures. Below is a preliminary procedure for implementing the gas buyback discussed in the previous section and in section 3.15 for initial pressurization of the gas export line. The major concern in this operation is temperature drop (JT effect) on pressure letdown during the initial phases of the operation.

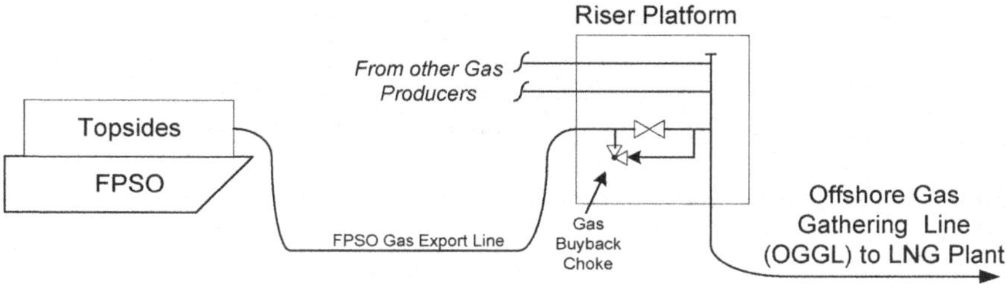

Success Scenario:

The steps in the process are:

<u>Prepare for Gas Buyback:</u>
1) Verify OGGL Gas is On Spec
2) Have LNG Plant Minimize the OGGL Header Pressure

<u>Initial Gas Buyback at Riser Platform:</u>
3) Initiate gas buyback (open buyback choke)
4) Build FPSO Gas Export Line pressure to 30 barg (out of extreme JT cooling region)
5) Release LNG to increase OGGL Header pressure to normal

<u>Purge Gas Export Line:</u>
6) Build FPSO Gas Export Line pressure to 70 barg. Operate Riser Platform to maintain pressure between 50 and 70 barg
7) Begin gas buyback at FPSO. Route to flare until nitrogen adequately purged.

<u>Purge and Pressurize Topsides</u>
8) Purge Topsides per Topsides Purging Procedure
9) Build FPSO Export Line pressure to maximum (100 barg +).
10) Close buyback valve at Riser Platform. Return Riser Platform system to normal state.

Potential Errors to Step 3

The table below contains extensions (potential errors) for one step (step 3) of the success scenario above. This structured approach identified many potential errors and allowed us to take steps to avoid those errors and/or to minimize their effects. For example, we identified choke failure as a significant risk since the choke had never been used before and the service was severe. We judged that we were not adequately prepared for this risk and secured additional spare parts.

Step	Potential Failure	Remedy / Remediation
3	Instrument failure – can't open buyback choke	Check prior to need, have instrument techs available
3	Instrument failure – can't monitor buyback gas temperature	Check prior to need, have instrument techs available
3	Lower than expected temperatures	Develop clear and unambiguous guidelines for temperature/pressure operating limits and operator responses to abnormalities and provide engineering support on Riser Platform
3	Loss of source flow during buyback (Riser Platform shutdown)	
3	Choke clogged or damaged by debris in line	Isolate, remove choke, clean and repair, reinstall. Identify and provide necessary spares, tools and resources to perform tasks.
3	Riser Platform services, power and utilities not available	Check ahead of time in time to repair
3	FPSO Gas Line system on Riser Platform not in full working order (this is the first use of the facility)	Validate commissioning effort. Checkout systems early enough to make any necessary corrections
3	Piping failure due to cold temperature	Establish effectiveness of shutdown system and develop procedures to insure personnel safety

5.7 IDENTIFYING POTENTIAL ERRORS IN IMPLEMENTING PROCEDURES

Accurate plans may be implemented incorrectly due to human error. In *Human Error* (28) James Reason identifies many potential causes of error. Below is a summary of his observations relevant to the potential for operator error in following a plan or procedure.

Guidelines for Evaluating Procedure Steps

Pre-conditions are likely to be incomplete or not done if/when:

1) Following maintenance/repair. Status of equipment after maintenance may not be as expected. For instance: slip blinds left in place, equipment not powered, safety systems bypassed, instruments on manual, etc.
2) Checklist does not follow a logical progression. For instance: valves in a line required to be in a given position should be listed in line order to facilitate checking.
3) Required conditions are ambiguous.

Errors in implementing steps in a procedure are most likely:

4) If a step is not obviously cued by the previous step it is likely to be skipped.
5) If there are too many steps, steps in the middle are likely to be skipped.
6) If a single step is complicated and has sub-steps, slips and lapses are likely.
7) A step which extends over long period of time without intervening action is prone to distraction. The operator may forget where he/she is in the process and/or fail to notice cues to continue to the next step.
8) Steps which occur after the main goal is achieved are likely to be skipped.
9) Interruptions from simultaneous operations cause you to lose your place.

Latent Errors – Problems waiting to happen

10) Look for latent errors caused by skipped steps. If a step can be skipped without impacting further steps or causing immediate upset to the process, then it is likely that failure to perform the step will not be caught.

Observations

This list is very useful for evaluation of process plant operating procedures. A procedure which is perfectly accurate may be difficult to follow accurately. This list helps identify areas where the operators are most likely to make mistakes.

5.8 THE PLANNING FALLACY – SENSEMAKING IN PLANNING

We frequently complete projects/tasks later than planned. This is partly caused by an optimistic bias in our planning. This optimism exists even among people who have previously performed similar tasks which took longer than they are predicting now and/or who have seen others perform similar tasks which took longer. Why are our plans consistently optimistic when judged against our results? Why don't we adjust our predictions over time to achieve more accurate schedule estimates?

There are several apparent causes of optimistic scheduling including:

- Plan-based forecasting is biased towards predicting success at every step. This simplistic planning fails to account for things that could go wrong.
- Early plans tend to be high level. High level plans tend to yield more optimistic schedules than more detailed plans. The project becomes anchored on optimistic high level plans.

- People tend not to use past experience, especially past failures, in predicting future performance. We do not expect to repeat failures.
- Early plans tend to be ambitious because they are generated early in the project when motivation is high. The schedule slips as motivation and enthusiasm slip with time.

Sensemaking – Success Bias

We have seen (chapter 2) that sensemaking simplifies history. We tend to view the past as much simpler and more deterministic than it really was. When an event occurs we can read past actions as leading inevitably to that result even if it would have been difficult to make those predictions before the fact.

We make the same errors in predicting the future. Our vision of the future is simpler than the future really will be. The actions we take will not lead inevitably to our desired result. The future may unfold in countless different ways that could impact the project. When we make sense of the future in this way we naturally view our tasks as simpler to accomplish than they really are.

High Level Plan Effect

High level plans tend to elicit optimistic schedule estimates. As we flesh out and add detail to action plans our estimates of duration become more realistic. But the early plans tend to be anchors which limit the impact of later, more detailed planning/scheduling efforts.

Failure to Use Historical Results

Perhaps the most surprising aspect of planning optimism is our failure to learn effectively from prior failures. If most of our projects finish late, then why don't we adjust our estimates to match our results? There are several possible reasons:

- When we are planning, our thoughts are on the future not on the past.
- History is not unambiguous. It is not always clear what happened on previous projects that caused them to be late. Different people evaluating the same project will identify different reasons for the problems experienced.
- When we can identify a reason why a previous project was late, we can argue that that particular problem won't happen this time.
- We tend to view our previous failures as being caused by bad luck rather than poor performance.

Less Bias in Predicting Other's Schedules

The scheduling bias tends to diminish when people are asked to predict how long it will take other people or other groups to do a task/project. That is, if we predict how long it will take us or our team to do a task we will be optimistic. If we predict how long it will take someone else or some other team to do the same task we tend to believe it will take them longer.

This probably results largely from an effect termed the self-serving attributional bias: our tendency to take credit for our successes and deny responsibility for our failures. When we succeed it is because we are skilled and competent; when we fail it is the result of bad luck. Conversely, we tend to attribute the failures of other teams (especially if they are competitors) to their incompetence and their successes to luck or unusual effort.

Sources

Observations in this section are based on several sources including the work on sensemaking by Weick (5), observations on the difficulty of using historical information by Neustadt and May (18), the study of social cognition by Fisk and Taylor (30) and especially on the discussion of the planning fallacy by Buehler, Griffin and Ross, Chapter 13 in *Heuristics and Biases* (29).

5.9 HUMAN ERROR MODELING

The past few sections deal with minimizing errors in action plans. Many of these insights come from the field of human error study. Over the next few sections we'll review the findings from that field. This information comes largely from Reason (28).

Errors Take a Limited Number of Forms

Actions proceed through three cognitive stages:
planning,
storage (of plan in memory) and
execution.

Errors take a limited number of forms and those forms are related to the stage in which they occur.

Cognitive Stage	Primary Error Type
Planning	Mistakes
Storage	Lapses
Execution	Slips

Slips and lapses result from attentional problems; typically from failure to pay attention at an important point in a largely automatic process. It is useful to view mistakes as occurring either in typically encountered situations in which we are applying an established rule or occurring in a novel situation. This results in the following error classification:

Skill-based errors:

Skill-based errors, slips and lapses, occur during the performance of automatic, unconsciously controlled behaviors, such as accidentally driving towards the office when you intend to go to the drug store.

Rule-based errors:

We solve many problems by applying rules. Rule-based errors involve applying the wrong rule to solve a problem. This typically occurs due to misreading the situation. An example is a baseball batter swinging too early at a changeup (mistaking it for a fastball).

Knowledge-based errors:

In novel situations actions must be planned on-line using conscious analytical processes and stored knowledge. Knowledge-based errors result from inadequate knowledge and from limitations of working memory.

Attentional and Automatic Subsystems

When you are learning to type you must pay conscious attention to each key you hit. This is a very slow process. As you become competent you learn to hit keys automatically, without conscious thought.

Cognitive activity occurs at two levels. We have an automatic subsystem whose activities are subconscious and an attentional (or conscious) subsystem. Most mental activity occurs in the automatic subsystem.

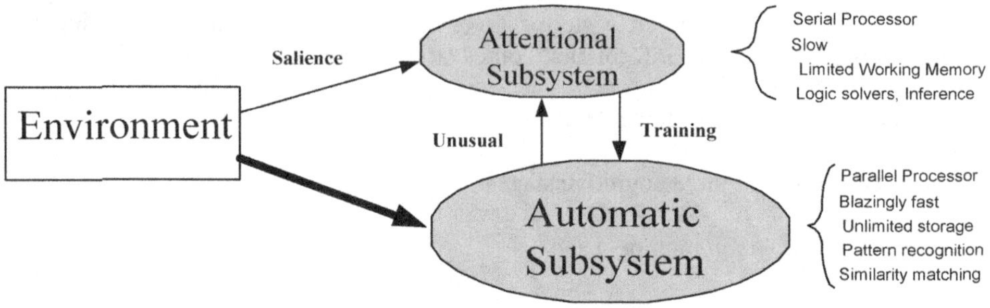

Automatic Subsystem: If you know how to type effectively it is because you have taught your automatic subsystem the rules of typing and that subsystem can now perform those rules without continually bothering your attentional (conscious) subsystem. The automatic subsystem is a huge parallel processor (capable of multiple simultaneous operations), is blazingly fast, and has almost unlimited storage capacity.

Attentional Subsystem: Issues which the automatic subsystem cannot handle or which are salient for some reason are raised into consciousness. Our attentional subsystem is computationally very powerful, but it is very slow and has very little memory (only a few items can be stored in short-term memory).

Similarity Matching and Frequency Gambling: Our automatic subsystem appears to function largely via two fundamental skills: similarity matching (pattern recognition) and frequency gambling. Both similarity matching and frequency estimation (how frequently we've seen a situation) appear to be performed quickly and effortlessly by our automatic subsystem. We'll consider these skills in more detail in our discussion of rule-based errors in section 5.11.

5.10 FAILURE MODES AT THE SKILL-BASED LEVEL

Skill-based actions are unconscious behaviors. These are well-learned behaviors that can, and sometimes must, be done without conscious attention. Typing, playing musical instruments, riding a bicycle, catching a baseball and many activities involved in driving a car are examples.

Errors at this level are related to errors in paying attention such as:

Double –capture slips: A routine action is performed instead of planned non-routine action. If you are driving to the drug store (uncommon) on the road that takes you to the office (common) you will drive on towards the office if you fail to pay attention at the right moment.

Omissions associated with interruptions: Interruptions can cause us to lose our place. The interruption may be counted as a step causing us to skip a step.

Reduced Intentionality: If an action is delayed the intention to perform it may be overlaid in conscious memory with other intentions.

Interference errors: Two active plans blend. I once put the cereal (instead of the milk) in the refrigerator while rushing to catch a carpool.

Over-attention: It is easy to see how lack of attention can cause slips and lapses. Paying attention to a process which is best left on autopilot also causes errors such as interpreting progress to be farther along or less far along than it is or seeing an expected cue that isn't there. Over-attention causes steps to be repeated or skipped. "Choking" by athletes is frequently caused by over-attention.

5.11 FAILURE MODES AT THE RULE-BASED LEVEL

Here we are talking about choosing the wrong rule.

Once we've learned to drive a car, most activities become learned skills. If the car is moving to the right we turn the wheel counter-clockwise to get back on course. This is a learned response; once learned we can do it automatically without conscious thought.

However, if the car is drifting to the right because it's sliding on ice there is a different rule. We are taught in driver's education class to steer into the slide. Since we are rarely in that situation we may not realize we are in a slide and even when we do realize it we may not apply the "steer into the slide" rule because we haven't learned it effectively. In this situation we are likely to apply the wrong rule.

Another example is 'driving on the wrong side of the road'. American who try to drive in a country where drivers used the left lane have significant difficulty acclimating to driving on the 'wrong' side of the road. Driving in the right lane is a habit, or a rule, that is not easily over-written.

Rules Compete

Rules compete for the right to be activated and applied. Success at being chosen depends on several factors including:

1) The stronger a rule is, the more likely it will be chosen. Rule strength is related to the number of times the rule has been applied successfully.
2) The more specifically a rule describes the current situation, the more likely it will be chosen.

Rule Hierarchies

Rules are arranged into hierarchies with general rules on top and more specific rules below. More specific rules are typically created when exceptions are noted in the applicability of the general rule (usually this occurs through failure of the general rule).

Similarity Matching and Frequency Gambling

Rule selection is enabled by two fundamental cognitive skills – similarity matching (pattern recognition) and frequency gambling. Similarity and frequency judgments are processed automatically without conscious effort.

Similarity Matching:	This extremely rapid and efficient cognitive process retrieves instances from memory which at least partially match the cues provided.
Frequency Gambling:	When cues under-specify a problem, several schemas may be partial fits. The essence of frequency gambling is that selection from among these rules is biased in favor of the ones most frequently used.

Causes of Errors in Applying Rules:

<u>Signs, countersigns and non-signs:</u> Situations which should evoke exceptions to a general rule often don't because of confusion or ambiguity of signs. Cues are often ambiguous. Signs will exist along with countersigns and often a great deal of noise (non-signs).

<u>Information overload:</u> An abundance of information makes it difficult to identify countersigns because we focus attention on the signs (information which in the past has been valuable in assessing the situation).

<u>Rule strength - Rigidity:</u> Rules gain strength by being used successfully. A strong rule is more likely to be selected. This is especially true if the cues are ambiguous. The tendency to use strong rules leads to confirmation bias – we see what we want to see.

<u>General Rules are likely to be stronger:</u> General rules are typically applied more often than more specific rules and hence become stronger.

<u>General vs. Specific Rules:</u> In unambiguous situations, people tend to use the lowest, most specific level of the rule hierarchy. But in practice, where cues are ambiguous, the tendency is to use more general rules (which tend to be stronger).

5.12 FAILURE MODES AT THE KNOWLEDGE-BASED LEVEL

Knowledge-based solutions are required when novel problems are faced. These are challenging situations. Success depends on having the knowledge or expertise to recognize the situation and define effective actions in real time. Dynamic system can be especially difficult to deal with. We are not very skilled at handling dynamic situations.

Causes of Knowledge-based Errors:

Selectivity: An important source of error is the selective processing of information. Mistakes occur if attention is given to the wrong features and/or not given to the right features. Our tendency is to focus attention on salient cues (bright, loud, large).

Short-term memory limitations: We are capable of maintaining only a few items in short-term (working) memory. In dynamic, stressful situations working memory is further limited. These memory limits constrain our ability to assess the situation and search for a solution.

Out of Sight, Out of Mind (availability): Studies suggest that while we are pretty good at identifying erroneous information we are poor at identifying missing data. When data important for assessing a situation is not readily provided we may not notice.

Confirmation Bias (sensemaking): Preliminary hypotheses formed with limited data interfere with future assessments. We tend to make sense of further data as fitting the pattern we identified initially.

Overconfidence: Problem solvers tend to be overconfident in their knowledge. They will tend to justify a planned course of action and discount contradictory signs.

5.13 SYSTEMS THEORY - UNINTENDED OUTCOMES

For every action there is an equal and opposite reaction.

This law applies to physical systems. It can't directly be applied to social systems, but something like it occurs in social systems as well. When you try to change systems they push back or adapt to the new forces in ways that may be difficult to anticipate. Often the net effect of a change actually is the opposite of the intended effect.

The use of radar for ship navigation is a good example. Radar was added to ships as a safety improvement. Used as intended, the addition of radar should have made ship operation significantly safer. However, it is not at all clear that radar has made ships safer. What is clear is that it has allowed ships to operate at higher speeds and under weather conditions that would have 'grounded' them in the past (pun intended). Ship operators have reacted to radar in a way that defeats the original goal of making shipping safer.

The Green Revolution in the mid-20th century increased crop yields dramatically. A given amount of land can now produce far more food than it did before. The result was expected to be a boon in agricultural profits and a decrease in hunger and famine world-wide. What actually happened is very different. Better seeds, fertilizers, pesticides and machines have made scale, capital and technology more important. This has led to the virtual extinction of the small family farm in the US. And the impact is much worse in developing countries. Nigeria is a good example; in 1970 Nigeria produced more food than it consumed. Today less ground is under cultivation and Nigeria imports half the food it consumes. Part of the reason for this is that the Green Revolution has priced most Nigerians out of the farming business just as it priced most small family farmers out of the business in the US.

More advanced medical technologies and ever more powerful drugs offer the promise of more healthy living. But increasing complexity means that medical professionals have to be more skilled. They now pay more attention to the results of specialized tests and focus less on the patients themselves. One effect is an increase in the frequency of errors by health care providers.

The levees around New Orleans provide excellent protection from floods that have plagued the city throughout its history. Until the middle of the 20th century houses in New Orleans were built off the ground on piers such that most floods did little damage. With the development of large and apparently invincible levees people began building houses on slabs at grade. When the levees failed due to Hurricane Katrina these slab houses flooded in many areas where raised houses were high and dry. The levees may have caused the flooding to be worse than it would have been had they never existed.

I could go on and on: low tar cigarettes encourage smoking, low fat products encourage dieters to eat more, computers increase the amount of paper used, etc.

Effective System Model

It is easy to predict that our actions will yield immediate success without side effects. We frequently fail to anticipate systems effects partly because our models are too simplistic. It was simplistic to believe that the only result of better seeds would be increased food production and decreased hunger. It was simplistic to believe that improved safety would be the only impact of radar on ships.

Both mechanical and social systems exist in a state of equilibrium with opposing forces in balance. In social systems actors have different and multiple objectives. When the system forces and resistances change, people will change their behaviors to optimize their situation.

To effectively predict the effect of any given action you must have a realistic model of the system. Where mechanical systems are involved the systems model includes physical forces and resistances. Where social systems are involved the model must account for the objectives, abilities and limitations of the stakeholders.

Example - Specialists on Projects

One area where use of simplistic models impacts project teams is the inclusion of more and more specialists on projects. The effect of this should be better decisions in the specialist's areas of expertise.

But the system fights back. Adding specialists often confuses roles and responsibilities. Specialists are often not well integrated into the project team and hence don't understand all the project objectives. Often they don't have a good knowledge of the industry outside of their area of expertise. The more specialists we add, and the more specialized they are, the less likely it is that they will be incorporated effectively into the project and the more likely it is that they will make poor decisions.

5.14 THEORY OF ACTION – Espoused Theory

This is the most important concept in the course.

The discussion of theories of action in the next few sections is taken mainly from the work of Chris Argyris (1,9).

Defensiveness is a fact of life in group settings. Many of our best ideas, plans and actions are defeated by opposition from others. Often this opposition appears unreasonable and irrational. It is often unexpected. For years this sort of behavior was puzzling to me. Fortunately, it is possible to understand defensive behavior.

In order to understand defensive behaviors and social limits to action we must understand the concept of theories of action.

Definition: **Your theory of action is the set of rules, conscious and unconscious, that you use to design your actions in social situations to achieve your objectives.**

Different people have different theories of action, of course. And our individual theories are very detailed and specialized. But on a macro level, our theories of action are very similar. We are all much more alike than we are different. For instance, in business and personal dealings, most people claim that they try to be fair, to consider other people's points of view, and to seek valid and accurate data.

Typical Espoused Theory of Action:
Try to be fair and honest
Seek and value accurate and valid data
Involve other stakeholders in the decision making

5.15 THEORY OF ACTION – Theories in Use

The espoused theory of action in the previous section feels right to most of us. At least it is the way we would like to behave. But we don't always behave that way. Let's consider some situations in which we are likely to deviate from this path.

Project Going Badly

Suppose one of your projects is going badly. Your boss asks you to make a presentation to the management team so they can understand your problems and help you recover. Your level of trust in your boss and the rest of the management team will have a large impact on what you choose to say. But, in general, you are likely to:

- sugarcoat the problems,
- focus on your plans for recovery rather than on the problems,
- focus attention on the less embarrassing features of the project,
- blame the usual suspects (lack of information, poor client scope)

- avoid blaming other people unless it's necessary and hardly ever blame anyone in the room. When we do blame others we frequently provide some face-saving information – "we all make mistakes and we were all so busy."
- And you are likely to omit any embarrassing information you're sure the boss is not already aware of.

Colleague Screws Up

One of your colleagues, not a close friend, has delivered some work to you which you are to incorporate into a report. It is a little bit late and a little bit sloppy and requires some cleanup before you can deliver it to the client. Are you most likely to:

- Complain to the client that your colleague's work is sloppy.
- Complain to your colleague's boss.
- Complain to your boss.
- Complain to your colleague (who did the work) and ask him to fix it.
- Correct the work yourself and don't complain to anyone.

For most of us, the probability increases as you go down the list. It is extremely unlikely that you will tell the client. It is also not likely that you will complain to your boss or your colleague's boss unless this happens repeatedly and you are sick and tired of it. You may complain to your colleague and ask him to correct the work. If the changes are easy to make it is likely you'll just correct them yourself and move on.

Suppose your colleague is also a good friend; are you more or less likely to point out the sloppiness in his work? Why?

Sales Pitch

Final negotiations are coming up for a big contract. You've just discovered that your team's proposal is technically flawed. You are certain that the client will not have detected the flaw. There is no time to fix the proposal before the final presentation and you fear that that would look bad anyway. Are you most likely to:

- Withdraw the proposal (let your competitor win).
- Keep the proposal in but point out the error and promise to fix it during detailed design.
- Proceed with the presentation as is. Correct the error if you get the job.

Many (most?) people will present the proposal as it is and fix the flaw later.

5.16 MODEL I vs. MODEL II BEHAVIOR

There are some common themes in the scenarios in the previous section. All of the examples contain violations of our espoused theory of action. Most of us behave according to our espoused theories in routine, low threat encounters. It is easy to be fair, to seek valid information and to involve other stakeholders if nothing important is at stake. But in situations where we feel threatened, or where the outcome of a decision is very important to us, we take a different tact.

- It is easy to be fair when nothing important is at stake. If something important to us is at stake we want to win.

- We seek and provide valid and accurate data as long as nothing very important is at stake. But if something important is at stake we are likely to cherry pick the data that supports our position.
- We willingly involve other stakeholders in routine situations. But when something important is at stake we may involve only those who agree with us.
- In potentially embarrassing situation we seek to save face.
- We try to avoid upsetting other people.

Theory of Action in Use – Threatening Situations:
We try to be in control, try to win
We seek data that supports our objectives
We act unilaterally in ways that will avoid inquiry
We try to save face and try to avoid negative feelings

The theory of action we use when threatened has been labeled Model I behavior by Argyris (also referred to as "theory in use").

The espoused theory has been labeled Model II behavior.

Model II – Espoused Theory	Model I – Theory-in-Use
non-threatening situations not much at stake	threatening or important situations great deal at stake
• **Try to be fair**	• **Try to win, to be in control**
• **Seek and value valid data**	• **Seek data that supports your objectives** • **Try to save face, avoid negative feeling**
• **Involve other stakeholders in decision making**	• **Act unilaterally in ways that inhibit inquiry** • **Involve only those stakeholders who agree with us**

The More Important the Situation, the Worse we Behave

Model I behavior kicks in when the outcome is important. We are more than willing to be fair when nothing important is at stake, but if what's at stake is very important to us, we want to be in control. We are at our worst in the most important situations. The cost of such behavior in the business world and in social settings is very high.

5.17 RIGHT-HAND LEFT-HAND EXERCISE

Pete is one of your subordinates. He has been turning out substandard reports lately. In your reviews of his work you've made the necessary comments to get them up to standards. His response to these comments has been brusque, bordering on unpleasant. It is time that you approached him to discuss the quality of his work. You are apprehensive because you believe he may react poorly.

On the right hand side of the form below construct the conversation between you ("Boss") and Pete as you think it might go. Start with the first thing you would say to start the conversation; follow with what Pete might say in response. Use the exact words of the conversation, not an explanation of how you would approach the conversation. Include at least three verbal exchanges by each of you.

When you have finished crafting the conversation fill in the left hand side of the form with what you were thinking at each step of the conversation, but did not say to Pete. Also try to imagine what Pete was thinking that he did not say and write those down.

Things thought but not said	**Conversation**
Boss:	Boss:
Pete:	Pete:
Boss:	Boss:
Pete:	Pete:
Boss:	Boss:
Pete:	Pete:
Bose:	Bose:

5.18 EVALUATING THE RIGHT HAND-LEFT HAND EXERCISE

TYPICAL EXERCISE RESULTS

Things thought but not said	Conversation
Boss: I'll start gently and maybe we can have a meaningful conversation. I can't tell him that the reports were complete crap, but he knows that anyway.	Boss: Pete, we need to talk. You've done great work for us in the past, but your last two reports weren't the quality that I know you are capable of.
Pete: I knew this was coming. If we got any real direction around here we could do more reasonable work. Besides, what's the point in writing a professional report when you'll change everything in it anyway?	Pete: Yeah, I guess I didn't put quite enough effort into them.
Boss: Didn't put enough effort! What a copout. He hardly did anything. He refuses my offers to help. How can I get him to admit that he needs to shape up so we can start working on the real issues?	Boss: Let's talk about that. Why aren't you putting in your usual effort?
Pete: He's right. I'm not putting in the effort. I have to work harder or quit. If he wasn't such a control freak. It used to be fun to work here.	Pete: Well it was just two reports; cut me a little slack here. I'll pick it up.
Boss: There he goes again; being defensive and we haven't even begun the conversation. What's the point in having a discussion? I'm sure he thinks my demands are totally unreasonable. Is my management style too overbearing? I could help him if he let me.	Boss: I'm sure you will. If there's anything I can do to help, let me know.
Pete: You could help me by transferring me to another group.	Pete: Thanks, I sure will.

<u>How did you start the conversation?</u>

If you have something negative to say about someone you will typically try not to upset the other person. You will likely try either or both of the following:

- using non-directive questioning and/or
- using face-saving approaches (easing in/diplomacy)

Did you start by complimenting Pete on some previous work?
Did you start with a question about whether something is bothering him?

<u>Why did you start that way rather than getting straight to the point?</u>

Do you think Pete is more likely to respond positively if you take this approach?

How did Pete respond?

Pete knows what you meant. He is probably embarrassed by the last two reports. He wants to get out of this discussion as painlessly as possible. So, he plays it coy, admitting as little as possible ("didn't put enough effort into it"). Saving face is probably his main goal. Avoiding an ugly, emotional discussion with his boss is another. He will blame the organization or you for his trouble if necessary as the discussion progresses, but he'd rather avoid it.

As the discussion progresses, note how counterproductive the strategies are. The words do not reflect what either the boss or Pete are thinking. But both guess pretty effectively what the other is thinking.

We assume that Pete and his boss would both espouse the Model II theory of action discussed above. Clearly, their behavior is not compatible with their espoused theories of action. After the session, both will feel that the other was being defensive and that nothing will change until the other person changes.

We carry on two conversations, a public dialogue and a private dialogue. The public dialogue (what we say) is carefully crafted to avoid upsetting others and maximize the odds of getting what we want. Our private dialogue (what we actually think), we keep private because we think it will upset others and decrease our odds of success. If we say what we think, we believe it will only make things worse. Yet these tactics are usually understood for what they are by others. If both parties are playing this game

- Communication suffers,
- We are unable to make effective decision because we lack the necessary data,
- We become defensive, and
- We make assumptions about the motives of others.
- Neither party checks their assumptions because their assumptions are obviously true and bringing them up would only make matters worse.

Solution:

Change your thoughts into something that can be made public and then say what you are thinking!

In the exercise above the Boss can't say what he's thinking because he's thinking that Pete is lazy and doesn't respect him. If she introspected on those thoughts she would likely come to a different point of view. Pete used to do good work. He's been very defensive and has recoiled from direction lately. This has caused the Boss to try to exert even more control so that the situation is now escalating out of control. Could the solution be to give Pete more freedom?

Could the boss conduct a more meaningful discussion with Pete if she approached the conversation with those thoughts in mind?

5.19 EMIGRES ENJOY FREEDOM TO PRAY – AND TO WORRY
Article in USA Today, July 11, 2000

Exercise: Evaluate the article on the next page for examples of Model I and Model II Behavior

The newspaper article on the next page appeared in USA today on July 11, 2000. Read the article and answer the questions below. In your answers evaluate the behaviors of the Communists, the Evangelicals, and the Orthodox Church as Model I vs. Model II behaviors.

Why did the Communists persecute religions in general and the Evangelicals in particular?

How did the Orthodox Church behave towards the Evangelicals after the Communists lost control?

What is the response of the emigrants to religious freedom in the U.S.? Is religious freedom their real goal?

Why are these people behaving in these ways?

Intentional power grab? or

Unintentional consequences of well-meaning people behaving in ways they are socially conditioned to behave?

Thousands of Evangelical Christians Settle in Pacific Northwest

By Martin Kasindorf
USA Today

PORTLAND, Ore. – Few Americans know that 50,000 Russian and Ukrainian Christian fundamentalists are living in this Pacific Northwest city as refugees from post-Soviet, small-town intolerance. Yelena Sergeyeva, 54, a former hotel manager from St. Petersburg, didn't know it either until she emigrated here eight years ago.

In fact, she says, she had little awareness that evangelical Christians existed in her homeland until she got a job supervising English-language classes and other refugee services funded by American churches.

Baptists, Pentacostals and other evangelical Protestants had been secretive in Soviet villages for decades – and with good reason. The Communist regime discouraged religion and jailed evangelical pastors and allowed the traditional Russian and Ukranian Orthodox churches to function only under tight control. State commisars, furious when evangelicals' children refused to join Communist youth brigades, denied them schooling.

Matters improved only briefly after the Soviet Union broke up. "Freedom for other religions was experienced at the beginning of the '90's, but it is not the same now," says the Rev. Peter Kryzchko, 41, senior pastor, of the 600-member Portland Slavic Evangelical Baptist Church. "The Orthodox religion prevails. There is less and less freedom, more persecution toward Pentacostals and evangelicals."

Guided by patriarchal ministers, entire born-again congregations packed up, flew to Canada and then applied for U.S. asylum. The refugees settled in Portland, Seattle and Sacramento. American churches helped with sponsorship and money. In 1992, Lutherans and the Mennonite Brotherhood lent $272,000 to the Rev. Vasiliy Vakulskiy, a refugee from western Ukraine. He bought a vacant brick church for his Slavic Pentacostal Church in southeastern Portland. The congregation has grown to 1500.

Refugee pastors say they have found religious freedom. But the American definition of religious freedom as state-neutral equality for all faiths mystifies them.

Its "not normal" that Christianity isn't more privileged than other religions,says Kryzhko, who arrived from Rostov, Russia, two years ago. Christian prayer should be mandatory in public schools, he says, and its "no excuse" to bar prayer because Buddhists or Muslims would be offended. "Nobody stops them from using their church and praying everywhere they want", he says.

The evangelical church preaches a hard-line social message that warns the young against American wickedness. Anna Bodnar, 23, a member of the Church of the Rebirth, warily arrived from a small Ukrainian town five months ago. "I expected so much freedom, like wild clothes and people without clothes, naked, a lot of sex stuff," she says. " I was kind of surprised it was not as much as I expected. It was kind of normal."

The immigrant churches would like to run their own parochial schools but can't yet afford it. Meanwhile, in public schools, the children pick up ideas that clash with church teachings.

"There is tension because parents don't allow children to dance," says Sereyeva, refugee program coordinator for Russian Oregon Social Services, an organization financed by Ecumenical Ministries of Oregon. "The kids would like to look like American kids. Their parents force them to go to church every night, force girls to wear long skirts and scarves. They are between two cultures.

America allows "too much freedom for the children" Vakulskiy says sternly. "In Russia, they wore uniforms, clean and proper-looking. Here, children go to school as if they were ready to go to the pasture to tend the cows."

The suburban David Douglas School District, which has 600 Russian-speaking students, has hired Garlina Kogan, a translation-service owner originally from Kiev, Ukraine, as a go-between with evangelical families. Kogan says the parents, accustomed to keeping their heads down, frown on placing their kids in gifted programs. "I don't want him to go too far," Kogan says a mother told her. "I want him to be like everybody else."

Refugee mothers stay at home tending family broods of as many as 12 children. The fathers, hampered by poor English, gradually have adapted to the job market. Some have opened restaurants. Some work in Portland's "Silicon Forest" high-tech assembly plants. The evangelicals are "low-education people, but they are very honest and hard-working, ready for any job," Sergeyeva says.

Dedicated to the USA

For all their misgivings about the American ways, the immigrants are dedicated to staying in the USA. "Jesus see we lived very hard, and he gave America for us to live better," says Tatyana Boshko, 62, a Baptist who came to Portland from Modlova in 1997 with her husband, Vladimir, 62. Their English needs polishing, and husband and wife are in classes at All Saints Episcopal Church to improve it.

"We must study English to be citizens," Tatyana says. "Oh! I today would like to be citizen."

5.20 EXERCISE RESULTS

Summary of Article:

- 50000 Russian and Ukrainian Christian fundamentalists are living in Portland.
- Evangelical Protestants had been secretive in Soviet villages for years. The Communist regime discouraged religion, jailed evangelical pastors, and allowed the Orthodox Churches to function only under tight control.
- Matters improved briefly after the Soviet Union broke up. According to Rev. Peter Kryzhko, "Freedom for other religions was experienced at the beginning of the '90's, but it is not the same now. The Orthodox Religion prevails. There is less and less freedom, more persecution towards Pentecostals and Evangelicals."
- Guided by patriarchal ministers, entire born-again congregations packed up, flew to Canada, and then applied for U.S. asylum. Refugee pastors say they have found religious freedom. But the American definition of religious freedom as state-neutral equality for all faiths mystifies them. It is "not normal" that Christianity isn't more privileged than other religions, says Kryzhko. Kryzhko argues that Christian prayer should be mandatory in public schools.

Evaluation:

<u>Why did the Communists persecute religions in general and the Evangelicals in particular?</u>

Religions were a threat to their absolute control. Fundamentalists, with their deep faith are harder to control and therefore represented a greater threat. They were probably also much easier to persecute because there were fewer of them. The Communists behavior is easily understood as Model I behavior.

<u>How did the Orthodox Church behave towards the Evangelicals after the Communists lost control?</u>

According to the Evangelicals they were oppressed worse by the Orthodox Church than by the Communists. If this is true, we can picture the Orthodox Church as trying to establish control. Though not a very religious attitude, this is understandable and predictable as Model I behavior.

Another explanation consistent with Model I behavior is possible. It may be that the Evangelicals expected too much from their newfound freedom. If they responded to religious freedom by attempting to exert their own control, then we would expect the Orthodox Church to respond in kind.

<u>What is the response of the emigrants to religious freedom in the U.S.? Is religious freedom their real goal?</u>

Borrowing a phrase from *Animal Farm*, they are disappointed at not being "more equal" than other religions. Indeed, it is clear that their dream is not religious freedom at all. Their dream is domination of other religions. They want to be in control – Model I behavior.

<u>Why are these people behaving in these ways?</u>

These are probably well-meaning people simply behaving the way they have been socially conditioned to behave and completely unaware of the contradictions between their espoused beliefs and their behaviors.

5.21 EXERCISE – REFLECTING ON GROUP PERFORMANCE

This story is taken from Argyris (1).

Situation: A consulting group has done substandard work on a study. The company had successfully completed a similar study for the same client previously. A strong case team had been put together including three managers from the company and two outside consultants. But the study failed miserably.

A Vice President (VP) of the company is conducting an after-action review meeting to determine what went wrong and to try to learn from the experience. He asks everyone to be candid so that the company can learn from the experience. The participants agree.

Summarized Conversation:

Mgr 1: We had lots of chiefs, but no Indians. The internal team organization was never made explicit. Ambiguity in the team caused resentment. And client conditions changed during the study.

Mgr 2: Everyone had his individual piece and no one knew what anyone else was doing.

Mgr 1: There was backbiting. The VP changed commitments during the study (and disappeared). Mgr 3 was supposed to compensate for losing the VP, but that was not made explicit.

Cons 1: We went into the field to collect data too early. Then when things changed we could not go back into the field cost effectively.

Mgr 2: The order was to go into the field to do market interviews. No one was in control. We were out of control.

Interviewer: When did you sense the project was out of control?

Mgr 2: Between the start of the project and the first presentation to the client. We rarely met as a case team without the client present.

Mgr 1: We needed you (speaking to VP) there during the first client presentation because we were getting nit-picked to death. We killed ourselves for a week preparing, only to have the presentation nit-picked to hell for nonsensical reasons.

Interviewer: What could the VP have done?

Mgr 1: Act Vice-Presidential! I remember sitting there saying to myself, you have three choices – keep your mouth shut, try to contribute, or yell "you're being idiots!"

Mgr 2: The VP could have pointed out that it was an interim presentation and that the level of detail they were seeking at that stage was getting us off track.

Mgr 3: I tried several times to do something, but there were so many levels of the client's organization there that it was a zoo. A VP could have pulled it off.

Cons 2: I didn't know how to pitch in because I didn't know what we were doing as a whole.

Mgr 3: I agree with most of what's been said. I probably should have been more forceful. After the first meeting we decided that they were crazy, not us, and that we would give them value for the money in the final analysis. There was an incredible fixation to come up with something special.

5.22 EXERCISE QUESTIONS– GROUP PERFORMANCE

1. **List causes of the problems experienced by the group during the study.**

2. **List some ideas to prevent recurrence.**

3. **Do you think the participants in this discussion behaved per the Model I (In-Use) or Model II (Espoused) theory of action when they were conducting the study?**

4. **Do you think the participants in this discussion behaved per the Model I (In-Use) or Model II (Espoused) theory of action during the after-action review meeting?**

5. **If they behaved per Model I during both the project <u>and</u> during the after-action review meeting, is it likely that they shared the true causes of the study failure?**

5.23 EXERCISE RESULTS – GROUP PERFORMANCE

The two lists below were generated by the actual participants in the study as the problem causes and suggestions for improvement.

Participants Assessment of Problem Causes

- Too many chiefs, not enough Indians
- Poor initial planning, field work started too early
- Team meetings held in presence of client
- No leadership, VP busy with other project
- Nit-picking client with changing objectives

Ideas for Improvement?

- Procedures for kickoff meetings and team meetings
- Procedures to clarify team hierarchies and reporting responsibilities
- VP actively involved in all projects
- VP attends initial presentations to client

How well do they match your lists?

Do you think these are the root causes of the problems?

Will their ideas for improvement have the desired effects? Will yours?

It seems pretty clear that these participants engaged in Model I behavior both during the study and during the after-action review. This is discussed in the next section.

5.24 GROUP PERFORMANCE EXERCISE – A DIFFERENT STORY

If we assume that people behave according to Model I theory of action when threatened, then we can assume that they acted consistent with Model I both during the project and during the after-action review with the VP. If so, then they would have been less than candid during the after-action review meeting. We can use the action model to infer what may have really happened. We can do this by returning to some of their observations and making sense of them as Model I behaviors.

Consider "**we were in the field too early. Then when the clients situation changed we could not cost effectively go back into the field**."

But, if the client's situation changed, it would have been reasonable to go back into the field at client expense. It seems more likely that data collection was done poorly (wrong data collected for the wrong reasons) and they then tried to blame the client. Probably, data collection occurred before the client's objectives were determined. The team was action oriented; they had done this sort of study before; they believed that they knew what the client needed. They dove into the field work to impress the client with early progress and to be in control from the start. They took control of the project, but the client probably expected to be in control.

They knew they were in trouble when they realized they had collected irrelevant data, but they failed to correct the error by going back into the field. That would have required that they admit the error to their management and redo fieldwork at their expense. Going to management at that stage would have cost them control of the project and caused them to lose face.

Poor data collection resulted in client criticism and in the project being "out of control". Or perhaps the poor data collection was the result of them being out of control. Recall that they were out of control "**before the first client meeting**."

Participants knew that they were out of control, but did nothing to correct it. Each individual did his/her task without calling attention to the lack of a shared view to management. Surfacing these problems to management would have resulted in the team losing control of the project. Each tried to stay in control by doing his/her own job, and they avoided creating negative feelings by keeping quiet. They tried to solve the problems by working hard to find something that would rescue the study. After the fact they blamed their problems on lack of leadership.

Nit-picking client with changing objectives

They blamed the client for the disastrous first meeting (nit-picking preliminary results). This was probably face-saving; a cover-up of poor performance to that point. If this was indeed a difficult client, the problems of the first meeting would have created an opportune time to raise the issues with management. The team didn't do that. Later on they blamed the client for nit-picking and the VP for not being there. In that respect they may have been correct; had the VP been at the meeting he would have seen the mess they were in and may have acted to correct the problems.

Situation Model

A model of the project as it probably occurred is shown on the graphic below.

The problem likely began with the belief "we've done this before, we know what they need." Because of that belief, they inferred the client's objectives and headed to the field prior to fully understanding what the client wanted.

This resulted in them collecting irrelevant data, which led to client criticism, especially in the first presentation to the client, and contributed to them being "out of control." (Part of the client's criticism probably stemmed from the client feeling out of control.)

Defensive mechanisms kicked in. They blamed the client, blamed their own leadership (or lack thereof), tried hard to rescue the project by "finding something." Notice that in blaming their leadership they blamed someone (VP) who wasn't actually there working on the project rather than blame a colleague on the project. What they never did was try to correct the problem. They never discussed the problems honestly among themselves, never surfaced the problems to management, never reevaluated their fundamental assumptions or tried to correct their fundamental errors.

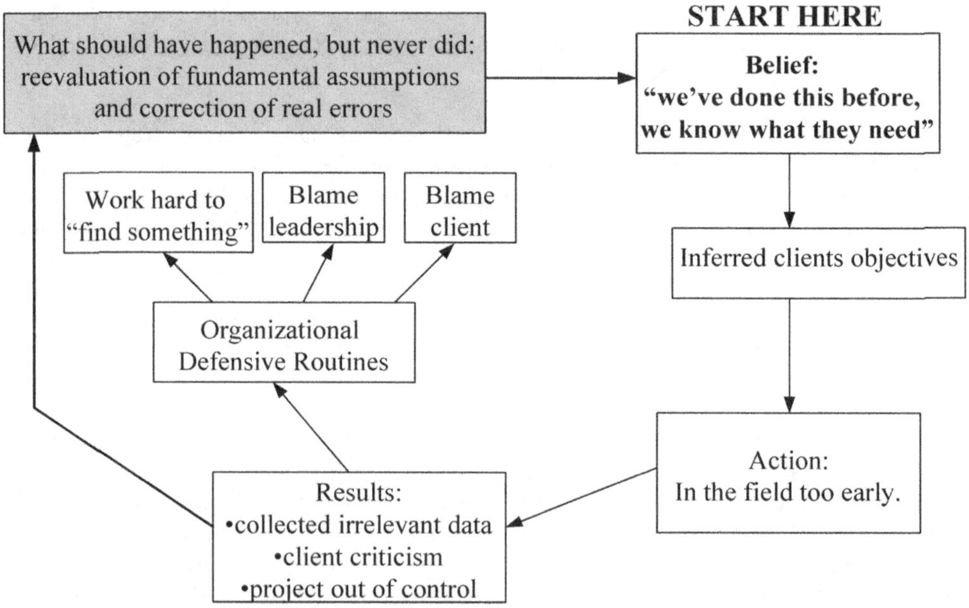

Model I Behavior Sabotaged the Project AND the Post Mortem Interview

All of their behaviors are consistent with Model I behavior. Model I behaviors sabotaged the project and also corrupted the results of the after-action review meeting so that little was learned at the meeting that would improve performance on future projects.

5.25 IMPACTS OF MODEL I BEHAVIOR

If we act in ways that violate our espoused theory of action then it follows that we will deny that we act that way: A typical pattern of behavior is:

- Espouse values
- Act in ways which conflict with espoused values
- Deny that a contradiction exists
- Make the subject undiscussible (and the undiscussibility undiscussible)

It is clear from research results that we do this largely unconsciously.

This behavior is perhaps most readily apparent among religious groups (some issues are intensely important to church members). It is also present in business. The impact of this behavior can be dramatic.

Typical Model I Behavior Pattern	Example of Model I Behavior in Religion	Example of Model I Behavior in Business
• Espouse Values	• Love your fellow man • Give glory to God	• People are our most important resource
• Actual behavior contradicts espoused values	• Strive to control lives and beliefs of others using any and all means including war	• Layoff in bad times, re-hire in good times • Minimal training budgets
• Deny that a contradiction exists	• Deny that contradiction exists	• Deny that contradiction exists
• Make discussion of contradiction undiscussible	• Heresy • Persecute dissenters (while proclaiming tolerance on all views and outwardly encouraging free and open discussion)	• Persecute dissenters ("not a team player", withhold promotions)

State Opinions as Fact – Limit Inquiry

In situations where we want to be in control a typical tactic is to state our opinions as though they are unquestionable fact. In this way we hope to inhibit inquiry and maximize our chance of winning.

5.26 REFLECTIVE PRACTICE

In *The Reflective Practitioner* (32) Donald Schon investigates the impact of Model I behavior on engineering practice. Schon views engineering as a conversation with the environment. When confronted with an existing situation that we want to improve we identify a possible improvement and then analyze how the change would work. As we analyze the idea the environment 'talks back to us', telling us whether the change will work as planned, whether there will be unexpected side effects, etc. The more changes we consider and the more effectively we consider them, the more back-talk we get, and the more we learn about the situation. Each move we make is an experiment. Both the problem definition and the solution evolve from this conversation.

Just as engineering practice involves an interactive conversation with the environment, so it also involves an interactive conversation with the client. But this conversation is limited by the theories of action of both the engineer and the client. The engineer is supposed to be confident, to understand the problem and to confidently develop a solution with minimal input. In addition the engineer will likely have preferences about how the problem should be solved. We then should expect the engineer to attempt to be in control, to seek data which supports her position, to fully develop her position before engaging the client in serious discussions (so that she will be in the best position to exhibit competence and to defend her solution).

But the client will also have preferences for how the problem should be solved and will wish to exert some control. In a situation where the engineer is playing it close to the vest the client feels left out and out of control.

The relationship is based on a degree of deception:

- The engineer betrays no sense of doubt or concern in her proposal to do the work. She believes such honesty would jeopardize her chances of getting the work.
- In confusing or ambiguous situations the engineer will cover up her confusion until she has had an opportunity to do further study.
- In situations where her initial efforts/ideas prove unsuccessful the engineer will stall while she checks out other possible solutions in order to present only the successful idea to the client.

The client observes the stonewalling and suspects lack of competence, exactly what the engineer is trying her best to avoid. Worse still, the client misses out on the fun part; the sense of discovery as the engineer converses with the problem. Deprived of this experience the client cannot have the same appreciation for the final design that the engineer does and cannot appreciate the tradeoffs the engineer had to make to arrive at the final solution.

The Alternative

The alternative, reflective practice, is to involve the client in the conversation with the environment; to expose the client to the ambiguity of the problem and to one's own uncertainties and concerns; to solve the problem together; to give up a measure of control in order to make the client a meaningful partner in developing the solution.

Consider how the case study we discussed in sections 5.21 to 5.24 could have turned out differently if the consultants had actively involved the client in the study rather than trying to control every aspect of the study themselves.

5.27 RELATIONSHIP BREAKDOWN IN THE WORKPLACE

An all too common phenomena in the workplace is for working relationships to deteriorate over time to the point where formerly cordial relationships become unproductive. We now have the tools to understand this relationship breakdown as a form of self-deception driven by Model I behavior and Sensemaking. Ideas for this section were taken largely from *Leadership and Self Deception* by the Arbinger Institute.

Initial Cordiality

With a new colleague you are likely to begin your relationship with a relatively trusting attitude and to give him the benefit of the doubt in early discussions. If this is so, initial discussions between the two of you will be relatively honest and straightforward.

If the initial discussions go well (you agree on substantive issues), then the level of trust will grow and further discussions will be more open and the extent and effectiveness of communication between the two of you will grow and improve.

Eventual Disagreement

Eventually, however, an issue will surface about which you disagree significantly. This is especially true if you represent different teams, companies, etc. If the issue is important to you, then you will try to control the situation in order to win (Model I behavior). In previous discussions, your communications were intended to inform. Now at least some of your communication is intended to control. In order to control the situation you will censor your conversation conveying only information which supports your point of view and you will largely ignore contradictory evidence. Your colleague will do the same. A wall begins to build between the two of you and conversation tends to deflect off the wall.

Sensemaking

In a perfect world the disagreement and disagreeable discussions would be forgotten and the next issue would be tackled in an atmosphere of trust so that disagreements would be only occasional events. Unfortunately, we do not behave that way. Once we have a disagreeable discussion we tend to make sense of the difference of opinion by making inferences about our 'adversary'. During initial discussion in which you agreed on substantive issues, you probably viewed him as reasonable, competent, friendly, easy to work with, etc. These traits do not seem consistent with his behavior when you disagree. We make sense of the situation by attributing traits to our adversary which would explain the difference of opinion; he now seems less competent, less intelligent, less reasonable, motivated by the wrong objectives, etc. This new view colors future discussions in insidious ways. Because of your new view your behavior will change. You've build a wall around yourself, as has he.

5.28 COGNITIVE BASIS FOR MODEL I BEHAVIOR

When researchers came to view Model I behavior as the cause of much workplace strife they set about trying to change people's behaviors. But efforts to get people to behave per Model II on a regular basis have proved unsuccessful. This has led researchers to conclude that Model I behavior is socialized into us at an early age and cannot easily be countered. No doubt that is true, but I also believe that Model I behavior is hard-wired into our brains and cannot be totally avoided.

Reptiles and Territory

What happens when you wander into the alligator's turf? It is likely to attack

Engineers and Turf

What happens when you wander onto the turf of a rotating equipment engineer? By turf I mean, of course, area of responsibility. Let's say you disagree with the design for a compressor control system and you change the P&ID's and bid documents without discussing this with the responsible engineer. Attack is likely!

Let's be clear that the attack will be about turf and not about substance; it will occur regardless of whether or not your solution is an improvement.

Structure of the Brain – Reptilian Complex

It is interesting to note that both behaviors can be traced to similar brain structures. Reptiles have much simpler brains than humans. Their instinctual behavior to protect their turf is guided by their simple brains. Nature has selected for aggressive alligators that defended their territory.

Human brains contain a similar construction. We have evolved higher cognitive functions which mitigate the instinctual aggressive behavior in many social settings, but the brain structure which drives reptilian behavior is still there and is activated in threatening situations.

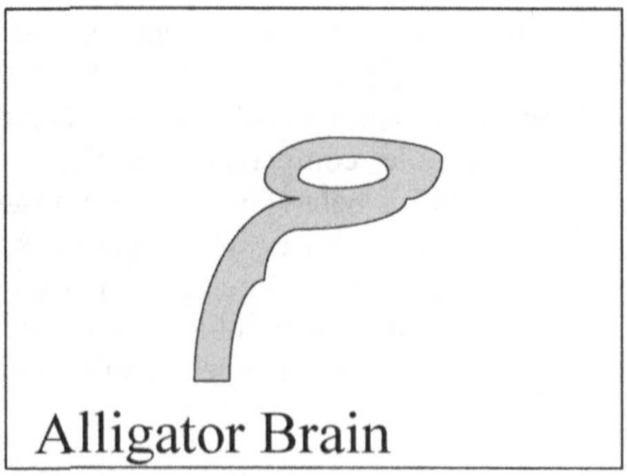

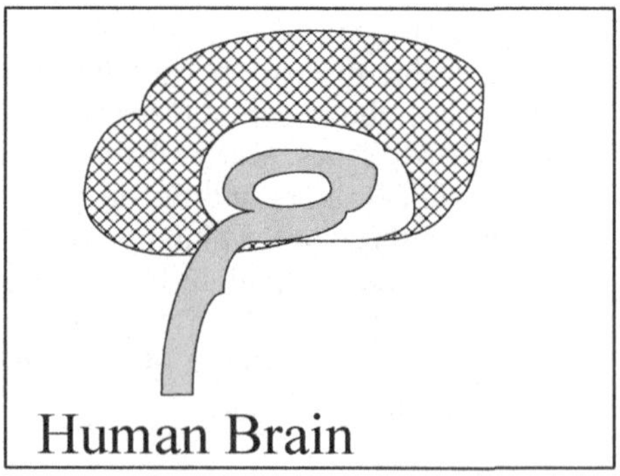

5.29 INFLUENCE

Social forces influence us, in certain situations, to do things that we would not normally do. Robert Cialdini identified many such situations in his book *Influences* (13). We'll discuss two of them here.

Reciprocity Rule - The Hare Krishna's Get Rich

The Hare Krishna's are a religious group with roots in India. They gained a footing in the US in the liberal 1970's. Their dress (flowing pink robes) and hairdo's (shaved head except for a ponytail) made them very recognizable. The fact that they were so easily identified made fundraising (panhandling in airports) difficult. Initially they were quite poor.

But then they hit on a gold mine. Either by luck or by excellent psychological insight they developed a remarkably effective panhandling approach. Rather than simply asking for donations, they began giving away paper flowers and/or books (The Bhagavad Gita).

They stressed that these were gifts (out of the kindness of their hearts) and did not accept "no" for an answer. Only after the gift had been graciously accepted would they ask for a small donation to their religion. This strategy increased donations dramatically. For a while, the Hare Krishna's were rolling in dough, building temples all over the US.

Why was this strategy so effective?

Argument:

Reciprocity Rule: When someone gives us something we are expected to respond back in kind. There is strong social pressure to reciprocate. No one likes a moocher. The rule applies even when the gift is not requested and not wanted.

Commitment Rule - Fraternity Hazing

Fraternity hazing can be brutal, degrading, painful, and dangerous. University administrations often try to ban them or control them. One interesting approach has been to try to get fraternities to substitute community service for the hazing ritual. This seems pretty reasonable – the community service could be made as challenging and demeaning as necessary. These attempts to control hazing are usually very ineffective. Hazing continues, even if they have to be done in secret.

Why are hazing rituals so important to fraternity organizations?

Argument

Commitment: We value things more if we have to work or suffer to get them. An effective model for this behavior is sensemaking. Having suffered the hazing you now have to make sense of why you did such a thing. The answer has to be that belonging to the club is worth it. The worse the hazing, the better the club looks. Fraternities know the importance of the hazing experience in building commitment.

Two More Theory of Action Rules:

What we have identified here are two additional rules in your theory of action:

- When someone gives you something you try to give something back.
- When you make a public commitment you stick to it.

5.30 PROTECTING YOURSELF IN A CROWD

Crowd Uncertainty

1964, Queens, NY: Catherine Genovese, a woman in her early 20's was brutally stalked and stabbed to death. The assailant attacked her three separate times over a 30 minute period. The first two times he was apparently scared away by the voices and bedroom lights of witnesses. Incredibly, 38 of her neighbors watched the attacks from the safety of their apartment windows. Not one of them came out to help. Not one of them even called the police until after she was dead.

Why didn't anyone come to help or at least call the police?

When an emergency is witnessed by only one person that person almost always stops to render aid. But when the same event is witnessed by many people, often no one offers aid.

The explanation has two parts:

<u>Shared Responsibility:</u> The behavior of a crowd is often governed by shared responsibility. When a single person witnesses an emergency, she knows that she alone is responsible and will almost always respond to provide help or call the police. When a crowd witnesses an emergency the responsibility to act is shared. It's possible no one will feel the responsibility enough to offer aid.

<u>Ambiguity:</u> Uncertainty plays a large role in crowd behavior. It is often difficult to determine what's going on in a crowd. You see a commotion, but is someone in trouble or is it just boisterous celebration? You aren't sure; so you look for the reactions of others. No one else is reacting, so you take that as evidence that everything is OK? But it may not be evidence at all. No one has actually made the judgment that everything is OK, everyone is simply going with the apparent judgment of the rest of the crowd and/or waiting for someone else to make a judgment.

Getting the Help You Need from a Crowd

We all seek safety in numbers in strange places. But, if the analysis above is correct, crowds may not be as safe as you think. Suppose you are on crowded subway platform and suddenly feel chest pains – is it a heart attack or just indigestion? You lean up against the wall. The feeling gets worse. You sit down. You moan in pain. People stare at you. You stare back. Your face may be saying "help me", but you don't voice the words out of fear of embarrassment.

The people who saw you sit down and who might have reason to believe you need help catch the next train. You pass out. Over the next few minutes dozens of people pass by assuming you are sleeping off a bad drunk.

How do you get the help you need?

- Ask for help while you still can. Don't wait until it's too late. Don't worry about being embarrassed.
- Single out **one** person. Ask that person directly for help. Make that **one** person responsible. Choose someone in authority (police, etc.) or someone visibly like you if possible (same age, sex, race, socio-economic group, etc.). Be specific "I think I'm having a heart attack, would you help me please?" Don't let there be any doubt in the person's mind that you need help.

Teamwork and Responsibility Ambiguity at Work

When it is not precisely clear whose responsibility a particular task is it may not get done even if everyone knows it needs to be done.

5.31 EMOTIONS IN DECISION MAKING

Over the next few sections we'll evaluate the role of emotions in decision making. Emotions play a dominant role in many social decisions. But it is a widely accepted belief that business decision making should be largely devoid of emotion. The story of Phineas Gage shoots some holes in that belief.

The Story of Phineas Gage (from Damasio (11))

In 1848, Phineas Gage, a railroad foreman, was tamping gunpowder into a hole. It was a standard operation. But this time a momentary lack of concentration resulted in a missed step in the process. A spark set off an explosion. The tamping rod, 1¼" in diameter and 3'7" long was sent flying through Gage's head. Gage did not die, indeed by all outward appearances he made a complete recovery and returned to work. But the Phineas Gage who returned to work was a changed man. Friends and associates reported that he was "a different person." He was unable or unwilling to perform his duties and was soon fired. He lived the rest of his life wandering from job to job.

Gage was not mentally impaired in the usual concepts of the word. His memory was intact. His reasoning was rational. He communicated effectively. In most respects he appeared competent and unchanged from the pre-accident Gage. But there was a dramatic change. <u>The post-accident Phineas Gage had a greatly diminished capacity to feel emotion.</u> Scientists have since identified areas of the brain which govern our feelings of emotion. One of these areas was damaged in Gage's brain. Mr. Gage had another disability as a result of the injury. <u>He was largely unable to make decisions in social settings.</u>

Emotions → Feeling & Caring → Ability to Decide

We discussed at length in Chapter 3 that caring is the reason we make decisions. Emotions are the root of caring. Phineas Gage could not make decisions because he didn't care. He didn't care because he did not feel emotions.

5.32 CAUSES OF EMOTIONS

Triggering Event

Emotions begin with a triggering event. These events might be real or imagined. You will feel anger if you think a friend has betrayed a confidence, even if she has not done so. The event may be something that causes harm to us or to someone we care about, or that increases the probability of harm. Or it may be a welcomed event such as a gift, or a favor that eliminates a threat of harm. Below is a list of some common emotions with possible triggering events.

Anger	Demeaning action against you or a loved one
Envy	Wanting what someone else has
Jealousy	Threat of loss of loved ones affection to another
Anxiety	Uncertain threat
Guilt	Having transgressed a moral imperative
Shame	Someone found out
Relief	Frustrating condition has changed for better
Hope	Fearing the worst but yearning for better
Sadness	Irrevocable loss
Happiness	Warm feeling of living a good life
Pride	Enhancement of self due to achievement
Love	???????
Gratitude	Appreciation of altruistic gift

From Lazarus and Lazarus (12)

Evaluation

Even a very harmful event will not cause emotional arousal if we don't evaluate it as a harmful event. Conversely, insignificant events that we evaluate to be significant do cause emotional arousal.

Personal Meaning

The product of evaluation is personal meaning. We have seen earlier in the course how meaning (belief) is generated and maintained.

Resulting Feelings and Objectives

If we care about the meaning we've assigned to an event we will have emotional feelings.

Action Tendency

A natural outcome of assigning personal meaning is the desire to change things we care about. Emotions tend to be accompanied by strong urges to take action. Anger is accompanied by the urge to attack; fear by the urge to flee; sadness by the urge to cry or for the company of friends, etc.

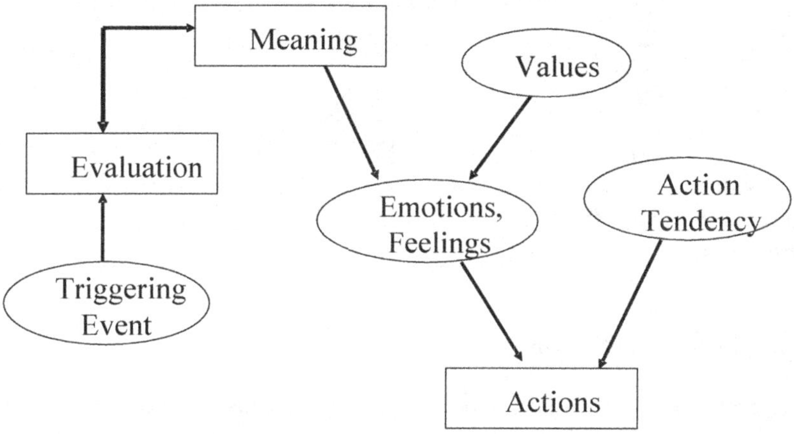

Analogous to the Decision Process

Compare the sketch above to the decision process we've been studying. They are analogous. Emotions and decision making are rooted in the same cognitive processes.

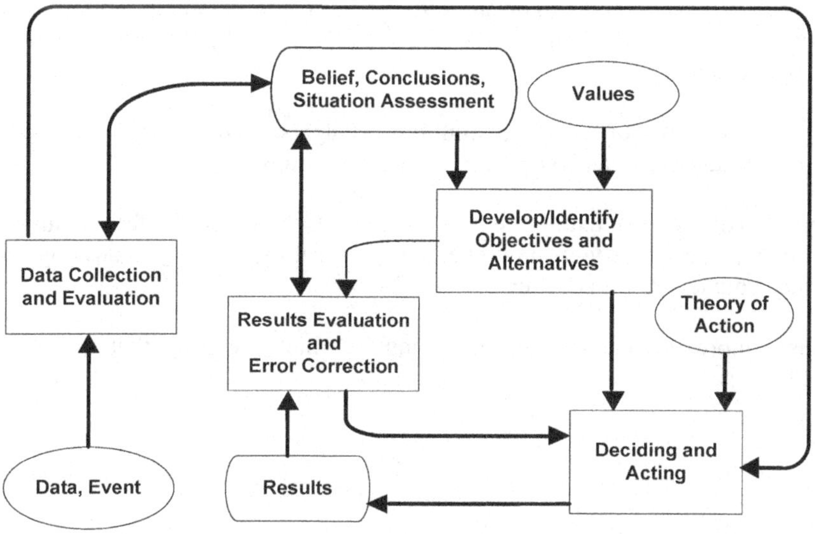

5.33 METHODS OF COPING WITH EMOTIONS

Coping with emotions can be action centered (problem solving) or emotion centered (dealing with the emotions themselves).

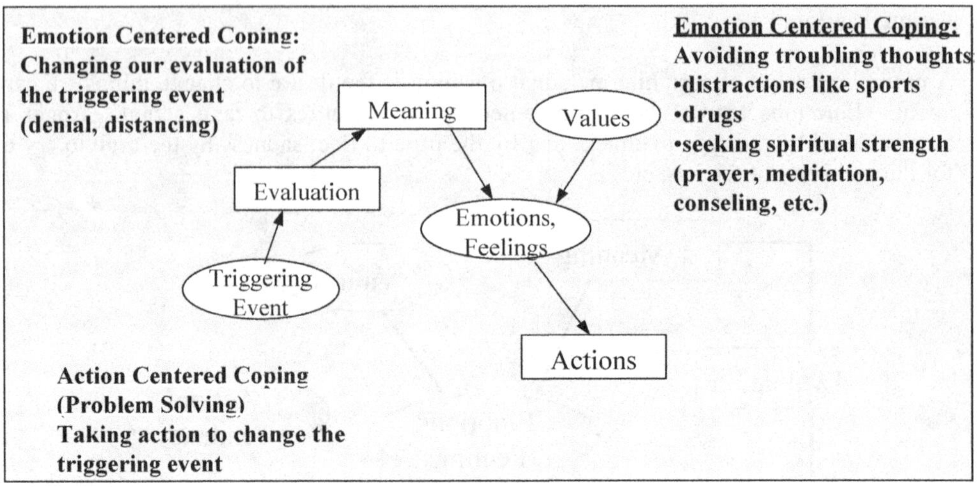

Action Centered (Problem Solving)

Emotions frequently move us to take action to correct the cause of the emotion. When we feel anger because of something that someone has said we may confront that person physically or verbally. Action which avenges our anger and/or causes the offending person to change his behavior will diminish the anger we feel.

Emotion Centered Coping

We do not always have access to meaningful action to confront emotions, and sometimes choose not to take action. Emotion centered coping can take the forms of distraction, distancing and denial.

Distractions are ways of avoiding troubling thoughts. Distractions can be healthy (sports, prayer) and they can be destructive (alcohol, drugs).

Distancing involves reevaluating the triggering event. You will not feel so hurt at being dumped by a loved one if you can argue that she wasn't good for you anyway. You try to convince yourself that it is better that it ended now.

Denial is a more dramatic means of coping in which we deny that the event even happened.

5.34 REFERENCE EFFECT – POSITIVE AND NEGATIVE EMOTIONS

It is interesting to compare the intensity of positive and negative emotions. Consider the two following scenarios.

1) Suppose you meet the wife of a business colleague and are instantly smitten. At several business functions you get to know her, fall in love and you consider trying to steal her away from him.
2) Suppose that you are married to the a wonderful woman. You've noticed that a business colleague has been paying an inordinate amount of attention to her at business function. You are convinced that he is trying to steal her away from you.

Compare the intensity of your feelings over these two scenarios. It is likely that the longing you would feel in scenario 1 is not nearly as intense as the jealousy you would feel in scenario 2.

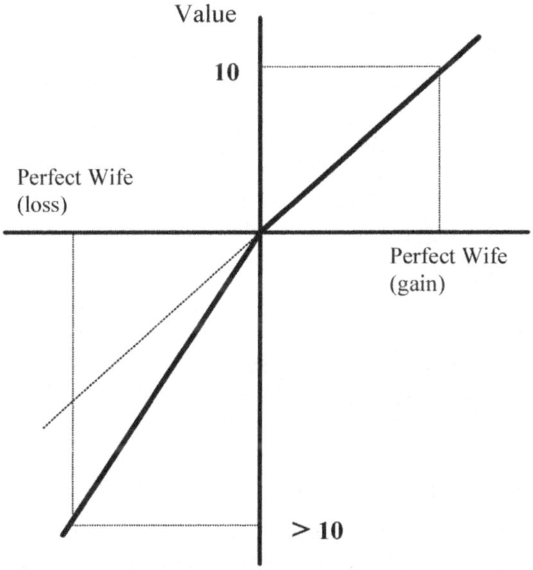

We should expect this behavior based on our earlier investigation of value curves. Longing is based on a potential gain; jealousy is based on a potential loss. Loss is felt more strongly than gain. Hence, we should expect the negative emotions to be more significant and more intense than the positive emotions:

Hate	>>	Love
Sadness	>>	Happiness
Despair	>>	Hope
Shame	>>	Pride

Does this suggest that the more risk averse you are, the more unhappy you are likely to be?

5.35 CHAPTER SUMMARY

We should seek alternatives that are both actionable and effective:

A plan may be unactionable if:

- you lack the physical skills and/or the knowledge to take the action.
- you lack the authority to effect actions that must be done by others (the action may still be actionable if you can convince them to do it).
- the required action requires communication skills, interpersonal skills or a level of self-confidence that you don't possess.
- the action item is too ambiguous, or is not formulated as an action.
- social limits to behavior make it difficult to implement the actions.

Actions are effective if they achieve our desired objectives without generating undesirable and unanticipated side effects. Ineffectiveness typically results from one or more of the following three causes:

- Errors
- Systems effects
- Defensiveness

Human error modeling provides useful insights into the causes of errors in implementing plans. Three types of errors are described:

- Skill-based errors
- Rule-based errors
- Knowledge-based errors

The most important part of this chapter and of this book is the discussion of social limits to action as described by the concept of theory of action. Two theories of action are described; Model II Behavior or Espoused Theory and Model I Behavior or Theory in use.

People claim to behave per Model II (Espoused theory). Features of Model II Behavior include:

- Try to be fair
- Seek and value valid data
- Involve stakeholders in decision making

But in important and threatening situations we tend to behave per Theory-in-use (Model I Behavior). Features of Model I Behavior include:

- Try to win (rather than be fair)
- Cherry pick data that supports your position (rather than valid data)
- Act unilaterally (rather than involving others)

Model I behavior often results in defensiveness and makes it difficult or impossible to implement some alternatives.

To assess whether a plan is actionable consider the following factors as determinants of how people might behave:

Control: If the plan is implemented who will gain control and who will lose control? People will fight for their turf however meaningless that turf might seem to others. This is especially true if they are surprised in public. Never underestimate this response. If you need someone to give up control to get what you want, be sure you secure their agreement in private.

Commitment: Have previous commitments been made? Will promises have to be broken?

Consistency: Have stakeholders already publicly expressed a view? If so, how difficult will it be for them to admit error? Can you provide a face-saving way out?

Leadership: Who is the leader? Who is in charge? Who will have energy about this issue and try to take charge? In an ambiguous or complex situation people will tend to follow someone who appears to understand the situation and has a strong opinion.

Emotion: All human behavior has an emotional component. If you expect people to behave in strictly rational ways you will frequently be surprised by their actual behaviors. Consider who will have an emotional attachment to a particular idea.

Wrong Behavior for the Right Reasons

Why do we behave according to Model I theory of action. It is easy to be cynical and view this as power grabbing, self-aggrandizing behavior. But usually people behave this way for the right reasons. People try to be in control when things are important to them because they want to see those things accomplished successfully.

Recognizing this behavior pattern in yourself and others can be a great first step towards solving workplace disagreements.

5.36 EVALUATING BUSINESS ADVICE (Including this course)

I argued in the first chapter that most business advice is bad advice and I promised to give you tools to evaluate business advice. We now have developed what we need. According to Argyris (1) there are three primary reasons why business advice fails:

1) Much of it is not actionable

Many business problems are caused by Model I behavior. A great deal of business advice proposes solving these problems by acting in ways consistent with Model II behavior. This works well for a while as people play nice. But Model II behavior is not sustainable in the long-term and people will revert to Model I behavior when important issues arise.

2) Much of it is ambiguous

A great deal of business advice is too ambiguous to be useful. Many suggested actions are not actions items at all. For instance, in *Wisdom of Teams*, Jon Katzenbach and Douglas Smith counsel us that building trust on teams only happens when people "do the hard work to build trust" – but nowhere in the book do they give any clues as to the nature of this work. They also encourage team leaders to "strike a balance between providing guidance and giving up control". But again, they provide no guidance on how one might go about determining what the appropriate balance is.

Ambiguous advice often sounds good until you go out and try to do it.

3) Business advice is frequently untested

No scientist would dream of publishing a new scientific theory without some experimental evidence to back it up, or at least a carefully reasoned argument. But commercially published business advice is rarely tested and often is not testable. Much of the advice given is appealing and compelling; often it appears to be obviously true. Because of the appearance of obvious truth, we don't insist on testing. Picking on *Wisdom of Teams* again, the authors claim to have "talked to hundreds of people on 50 teams" to develop their insights, but they don't reveal what they talked about and how those discussions led them to their conclusions. And they claim that "we have not statistically validated these findings." Imagine a scientist making such a statement in a published work.

I believe that the advice presented in this course passes these tests. I encourage you to apply the tests and judge for yourself and I would be eager to hear your thoughts if you find any bad advice.

5.37 ANSWERING THE FOUR QUESTIONS

What observations from this chapter can we make about our four key engineering questions?

Why do so many changes occur late in projects?

Planning errors discovered too late:

Preliminary plans are often high level and incomplete and may miss significant issues.

Difficult to admit error:

An error found in the design may not be immediately corrected if admitting the error would be embarrassing.

Commitment:

Once we commit publicly to a course of action it is difficult to change our position without losing face.

Design Completed before Details are Shared with Others

In order to remain in control and avoid embarrassment, engineers may tend to complete the design before communicating with others. The completed design becomes an anchor that is difficult to change.

Why do we have so many problems at interfaces?

Model I behavior:

Defensive behavior is accentuated by team membership. We are often more aggressive in defending the team than we are in defending ourselves.

Design Completed before Details are Shared with Others

In order to remain in control and avoid embarrassment, engineers may tend to complete the design before communicating with others. The completed design becomes an anchor that is difficult to change.

Ambiguity in Plans:

Ambiguous plans can often be managed within a cohesive group as group members routinely share data and knowledge. But when the relationships are more distant, ambiguity in plans may not be resolved consistently.

Responsibility Ambiguity

If it is not clear whose responsibility a task is it may not get done or it may be done by more than one person.

Why do we repeat past mistakes?

Inability to recognize unactionable plans:

A great deal of business advice recommends Model II behavior as a solution to problems created by Model I behavior. Since people are incapable of sustained Model II behavior the advice is unactionable.

Systems effects complicate learning:

Actions often have unintended side effects that undo the good that the actions were supposed to accomplish. It is difficult to recognize those side effects, especially if there is a time delay between the apparently positive outcomes which occur quickly and the negative effects which occur later.

Commitment:

For a variety of reasons people will make commitments to a course of action. If commitments have been made changing course can be difficult.

Why do we so often finish behind schedule?

Incomplete Plans:

Early plans that are the basis for initial schedule estimates may be incomplete. Plans which omit required tasks lead to optimistic schedules.

Unactionable Plans:

Plans that some stakeholders disagree with may be unactionable. Implementing the plans over their objections may be time consuming and may cause delays. Changing the plans may be time consuming and cause schedule delays.

High level plans yield optimistic estimates

Early plans tend to yield optimistic schedules.

Planning for Success

We tend to plan for success at each step in the process. But assuming that everything will go right is not realistic.

Chapter 6
Learning

Sources:

This chapter is based mainly on action science research, primarily the work of Chris Argyris. See *On Organizational Learning* (9). The discussion on organizational learning is largely from *Demystifying Organizational Learning* (34) by Lipshitz, Friedman and Popper.

Preface:

We are effective at learning unimportant things, but poor at learning threatening, important things. This chapter investigates factors that inhibit learning in social settings.

Prerequisite:

To understand this chapter you must thoroughly understand the concept of theories of action as discussed in sections 5.14 through 5.26.

Those who don't know their history are doomed to repeat it.

George Santayana

Those who do know their history also seem doomed to repeat it.

paraphrased from Neustadt and May, *Thinking in Time* (18)

The only thing we learn from history is that we don't learn anything from history.

George Bernard Shaw

CHAPTER 6 CONTENTS

6.1 SINGLE LOOP VS. DOUBLE LOOP LEARNING
6.2 SINGLE LOOP LEARNING
6.3 LOOP II – DOUBLE LOOP LEARNING
6.4 CASE STUDY – TAP
6.5 TAP CASE STUDY – GESTATION PERIOD
6.6 TAP CASE STUDY – EXPANSION PERIOD
6.7 TAP CASE STUDY – CONTRACTION PERIOD
6.8 EXERCISE – TAP CASE STUDY
6.9 TAP CASE STUDY EXERCISE RESULTS
6.10 SINGLE VS. DOUBLE LOOP LEARNING – TAP CASE STUDY
6.11 GRAND TETON DAM DISASTER
6.12 EXERCISE – GRAND TETON DAM
6.13 GRAND TETON DAM EXERCISE ANSWERS
6.14 GRAND TETON DAM – HRO PERSPECTIVE
6.15 LEARNING FROM COMPUTER PROGRAMMERS
6.16 ORGANIZATIONAL LEARNING
6.17 LEARNING THE OBVIOUS AND MISSING THE POINT
6.18 SUMMARY – FACTORS INHIBITING LEARNING
6.19 ANSWERING THE FOUR QUESTIONS

6.1 SINGLE LOOP VS. DOUBLE LOOP LEARNING

The final step of the decision making process is learning.

Definition:

Learning is any increase in knowledge or skill that enables the learner to be more effective at planning and implementing actions which achieve his/her objectives.

A lot of what we think of as learning isn't actually learning. Making a mistake and recognizing the mistake is not learning. We cannot be certain that we've learned something until we take action that successfully accomplishes our objectives.

Richard Farson makes the observation in *Management of the Absurd* (7) "we learn not from our failures, but from our successes." The common premise that we learn from our mistakes is flawed in two respects:

a) We may notice a problem, but not do anything to correct it. For instance, if we continually notice that silos decrease trust and cause poor performance in our organization, but don't take any effective steps to correct the problem; that's not learning.
b) Different people can learn different things from the same error. Some of the things we "learn" from mistakes will not successfully correct the problem and hence are not learnings at all. We can't be sure until we succeed.

System Theory

Learning the wrong lesson from mistakes often results from not taking a systems view. The "obvious" cause of a problem might not be the real cause or the whole cause.

Learning Loops

The graphic below shows two different leaning loops. The lower one (dashed lines and labeled "I") illustrates typical, trivial learning which results when we find a simple error in implementation of the decision that is easily corrected. The other loop (bolder lines) labeled "II" results when we learn that our fundamental beliefs about the problem are incorrect (which may have led us to identify invalid objectives and/or choose a poor alternative, etc.). Clearly this sort of learning is more difficult to accomplish.

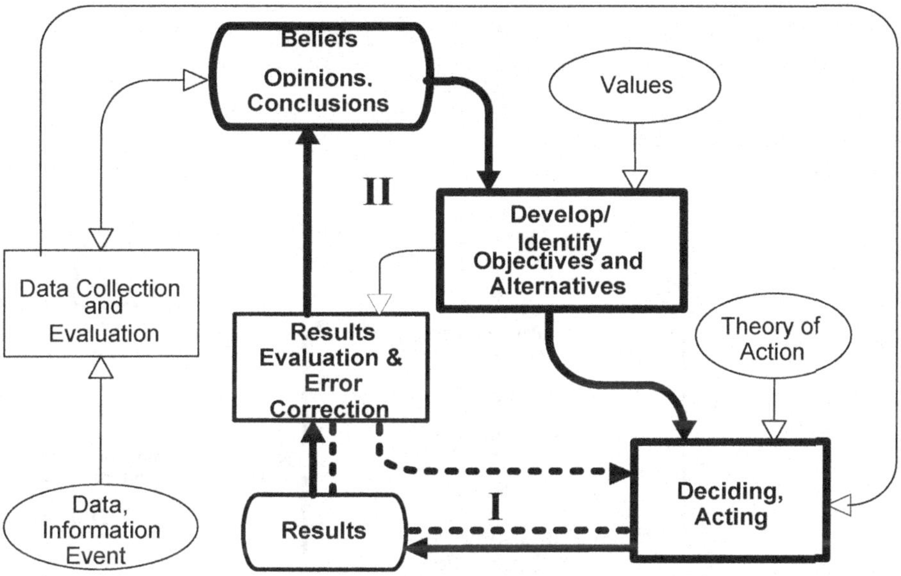

Argyris (9) identifies two types of learning which he refers to as single loop and double loop learning as described in the graphic below. Argyris' loops do not correspond exactly to the two loops above, but there is a relationship as we will see in the next two sections.

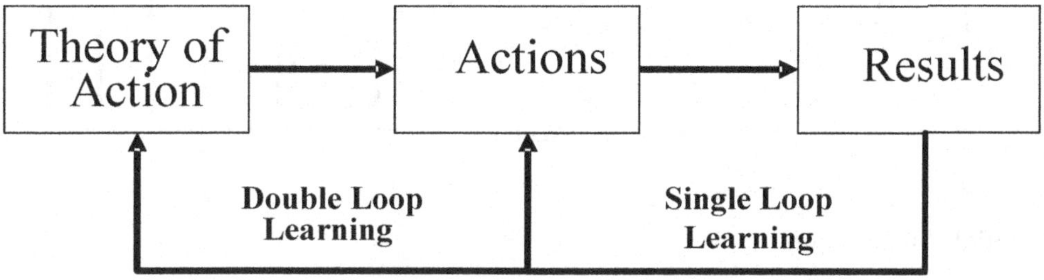

Single Loop Learning (I):

Often the results of our actions reveal a simple procedural error that is easily corrected. This is low stress learning. The fault is relatively easily corrected by changing our actions and the next attempt often goes more smoothly. Single loop learning is common and easy.

Double Loop Learning (II):

But what if the results of our actions are so inconsistent with our expectations that our assumptions and/or fundamental beliefs must be wrong? Learning such things is threatening and difficult. Model I behavior inhibits such learning. In order to learn threatening or embarrassing things we must overcome the limitations of Model I behavior. Double loop learning requires that we behave per Model II Theory of Action.

6.2 SINGLE LOOP LEARNING

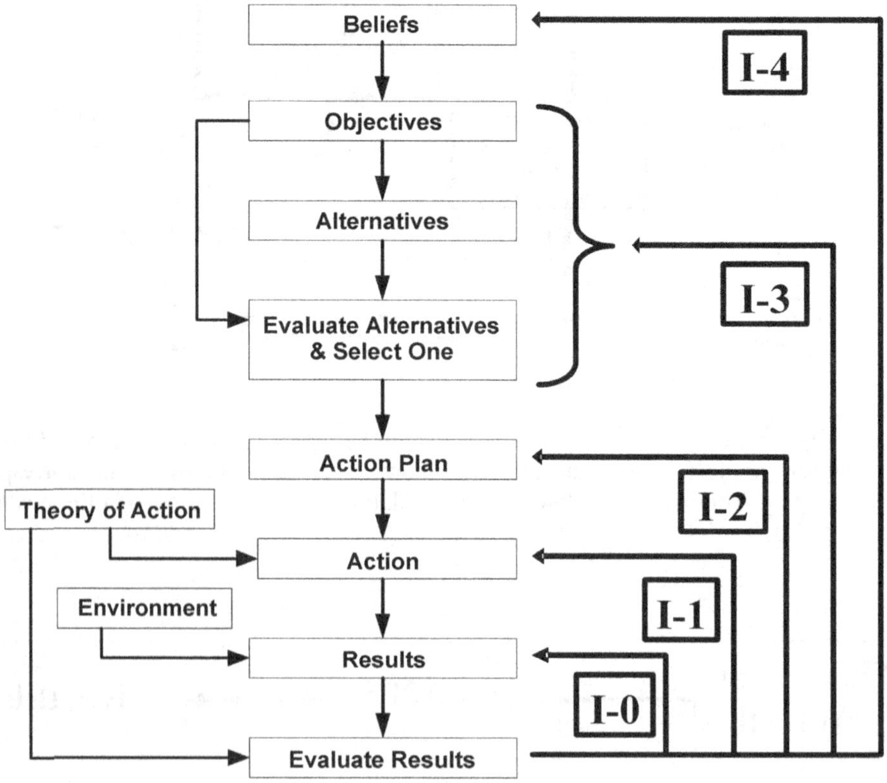

Multiple Single Loops

To fully characterize learning we need to talk about more than two loops. Single loop learning can take many forms. The figure above shows a decision trajectory from beliefs down to evaluation of the results. On the right are five possible learning results. These are all single loops.

Loop I-0 – Success or Bad Luck

Loop I-0 learning occurs when we are successful; the results match our expectations. We could also characterize it as loop I-0 learning when we conclude that poor results were simply due to bad luck; that if we did it all over again we would probably succeed next time.

Loop I-1 - Skill-based Error (flawed action)

Loop I-1 learning results when we determine that a flaw occurred in executing the plan (but that the plan itself is a good one). This could have been a skill-based error (slip or a lapse) or a rule-based mistake (misinterpreting a cue and applying an inappropriate rule). We judge that the plan is adequate.

Loop I-2 – Knowledge-based Error (flawed plan)

Loop I-2 learning occurs when we attribute failure to a flawed plan.

Loop I-3 - Judgment-based Error

Loop I-3 learning occurs when we decide that we chose a poor alternative. This could have resulted from a poor understanding of the objectives, a failure to identify promising alternatives or a flawed evaluation of the alternatives identified.

Loop I-4 - Belief-based Error

Loop I-4 learning occurs when we identify an error in our fundamental assumptions or beliefs.

Low Level Learning is Easy – High Level Learning is Hard

The fact that they are all single loops does not mean that they are all equally facile. It is easy to identify and correct slips and lapses and not that difficult to identify and correct rule-based errors. But the higher up the list we go, the more threatening and more embarrassing it is to identify or admit error. Whenever the results are ambiguous we will tend to favor a lower level explanation:

- tend to judge the results a success if possible,
- tend to blame failure on bad luck rather than error,
- tend to blame slips, lapses or rule-based errors rather than knowledge-based errors (flawed plan),
- tend to blame a flawed plan rather than judgment errors, and
- tend to blame judgment errors rather than flawed beliefs.

Placing Blame – When Someone else is Responsible

The discussion above assumed that you were responsible for all steps in the decision process. But often different people and teams are involved at different stages. Identifying error can be much less threatening if someone else made the error. If someone else created the action plan it is pretty easy to blame their plan rather than your implementation.

Example:

Suppose that you are the head of a special police task force whose mission is to decrease the use of the drug ecstasy. Over the past 3 years you have convinced the legislature to increase prison sentences for ecstasy use and have diligently pursued dealers and users, dramatically increasing the arrest rate and the conviction rate. Despite this, it is clear that ecstasy use has increased dramatically in your city. What are you likely to learn from this experience?

Loop	Learnings	Learning Difficulty
0	Could it be bad luck? You may point out that the consumption of ecstasy has increased nationwide and that your city is not exceptional. You may also point out that the cost of ecstasy has come way down with the increased market. You may even claim success by pointing out how many dealers are in jail and how bad things would be if you had not acted.	Easiest
1	Was your plan inadequately implemented? The results may have been more positive if police had spent more time and effort on the problem.	Easy
2	Was the plan adequate? Could the plan have been better thought out? Perhaps better results could have been achieved with more focus on major dealers instead of on local street dealers.	Hard
3	Another city in the state seems to be having positive results by focusing on treatment and education rather than prosecution. Could it be that police action isn't the most effective approach?	Harder
4	Some people are actually suggesting legalization! They point out that the drug actually has minimal detrimental effects; no worse than alcohol, and that prosecution takes away resources from more important police work.	Very Hard

6.3 LOOP II – DOUBLE LOOP LEARNING

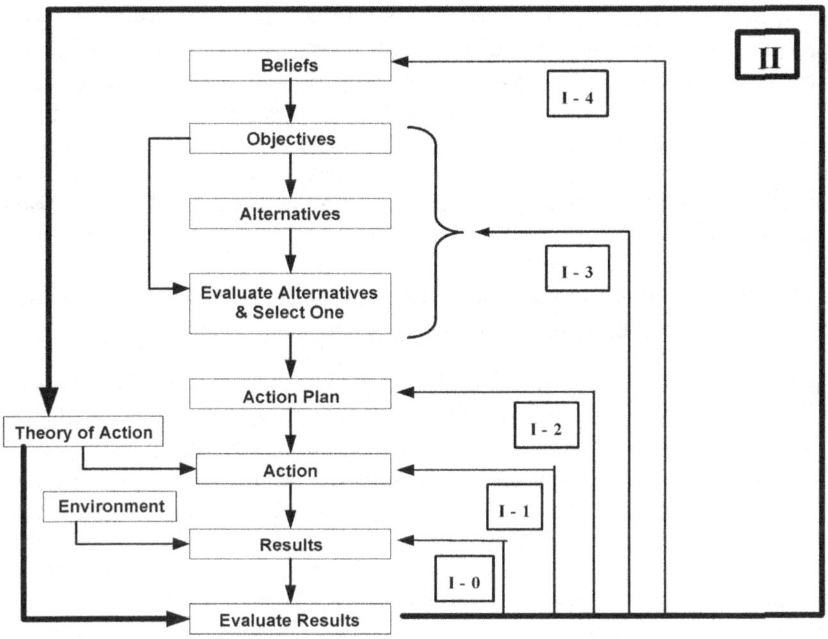

We tend to behave per Model I theory of action when evaluating errors. This will cause defensiveness and tend to inhibit learning at the higher levels. Double-loop learning requires Model II behavior.

The Loop labeled "II" above shows double loop learning. Here the actor recognizes that the situation is potentially embarrassing and that she must consciously introspect on her reasoning process in order to identify valid learnings.

Returning to the Ecstasy example, let's assume that the task force head has made an appraisal of the program and his initial assessment is that it was a good plan, expertly implemented and only bad luck (the increased use and price drop of Ecstasy nationwide) prevented success (Loop I-0 learning or effectively no learning). Later on he visits a city with much better results than his. That city has focused its efforts on education and drug treatment. Initially he rejects the results (they must be wrong), then he looks for extenuating circumstances (perhaps a different drug is more popular in that city). Only after much introspection will he possibly accept that his objectives and alternatives may have been misguided (level I-3 learning). This observation would have to be preceded by a fair and objective evaluation of threatening data (Model II behavior).

6.4 CASE STUDY – TAP

The Therapeutic Apheresis Technology (TAP) Case Study is an interesting story on the subject of trial and error learning. The story of TAP was first reported by Van de Ven and Polley and evaluated by Argyris (9). Van de Ven and Polley presented the TAP story as a study on adaptive learning or trial and error learning. The concept of trial and error learning is simple; that we persist in a course of action if we consider the results to be positive and try something else if we perceive the results to be negative.

TAP was a joint venture of three companies seeking to develop therapeutic apheresis technology. Apheresis is a filtering technology used currently to separate blood components. I occasionally donate blood platelets. In the process, blood is extracted from my arm; platelets are filtered out of the blood via apheresis technology, and the rest of the blood is returned to my arm. The TAP venture sought to improve this technology to the point where it could be used for treating disease by filtering pathogens out of the blood. The plan involved initial development and marketing of a product to compete with existing apheresis technology, followed by products designed to treat diseases using advanced filtration technology.

Van de Ven and Polley conducted a real-time study of TAP's development. They divided the venture's eight year life into three periods:

Gestation Period: In the initial period (gestation) researchers developed plans and conducted preliminary research to validate the plans. The researchers met regularly in this period with resource managers to present the results of their research and share future plans. The resource managers (project managers) controlled project funding and reported to executives of the three partners. The gestation period culminated in funds approval for product development.

Expansion Period: Following approval of funds the research effort was expanded and the team worked towards development of a commercial product. This phase ended with commercialization of the first TAP product.

Contraction Period: The initial product sold by TAP experienced problems in the field. Planned Phase 2 products were delayed. Eventually the project was terminated.

6.5 TAP CASE STUDY – GESTATION PERIOD

Researchers Made Optimistic Claims to Secure Funding

TAP researchers were enthusiastic about their jobs and the opportunity afforded them by the TAP venture. Their early research efforts were ambitious. They worked diligently to succeed. But the early results were not all they had hoped for. Still they persisted. Early failures were not explicitly divulged outside the research group. Status meetings with the resource managers were well planned, highly orchestrated events. Setbacks were sugar-coated. They maintained optimistic projections.

Throughout this phase, the researchers had doubts and fears. These fears were never made public outside the research team. They all believed that, with effort, things would get better.

Resource Managers Publicly Accepted Plans that they Privately Referred to as "Fluff"

The resource managers publicly accepted the sugar-coated results and optimistic projections as presented to them by the researchers and conveyed this optimistic image to the partner's executives. They did this even though privately they believed that the projections were largely "fluff". They believed there would be changes to the projections in the future.

Learning not related to Results

Van de Ven and Polley concluded that no trial and error learning occurred during the gestation period. Outcomes of experiments and future actions seemed unrelated. Research setbacks did not cause any change in plans.

6.6 TAP CASE STUDY – EXPANSION PERIOD

The gestation period ended with the approval of funds and was followed by the expansion period.

Setbacks occurred frequently in the expansion phase. The researchers were held accountable for meeting the overly optimistic plans developed in the gestation period. The pattern of impression management, begun in the gestation period, was continued in this phase. The researchers sugar coated their results, never reporting any uncertainty, or even suggesting that the development would take longer than planned.

While development efforts were largely successful in this period, there were significant problems with the manufacture and scale up of the filtration unit (the key technology!). Eventually these problems caused delay of introduction of the unit by several months. At administrative reviews, these problems were presented in ways designed to prevent inquiry and to discourage problem solving and learning. The researchers prepared slick visuals, rehearsed their presentations, prepared answers for anticipated questions, and framed negative information in the most positive possible light.

Most errors noted in this phase were dismissed by the researchers as irrelevant, if discussed at all.

Van de Ven and Polley noted that during the expansion phase many parallel and independent streams of research were conducted. This had the effect of distracting attention from the core innovative idea, the filtration system. Research on the filter itself was going poorly, but no one seemed to notice. Participants claimed that so much information was circulated it was impossible to keep on top of it all.

Problems were compounded when key personnel were reassigned to other projects.

Learning Negatively Correlated with Results

Van de Van and Polley observed a negative relation between outcomes and actions. Not only did experimental errors not lead to reevaluation of a failing course of action, failure seemed to make the researchers even more committed to their course.

6.7 TAP CASE STUDY – CONTRACTION PERIOD

Things changed when the first commercial product was introduced. Sales were poor and units that were sold experienced significant problems. The bad news was now public information and could not be hidden from management or sugar-coated away.

The resource managers finally began to ask serious questions and intervened to break the patterns of persevering on a failed course of action. Sugar-coated presentations were now challenged.

The researcher's focus also shifted towards identifying and solving real problems. Long-standing manufacturing problems were fixed and the TAP unit achieved reliability comparable to competitors units.

The researchers saw the setbacks as temporary. They believed they had turned the corner and the future was bright. But the resource managers had lost confidence in the researchers. The two parties developed alternative stories that interpreted the same events quite differently.

TAP was canceled just when it looked like things were finally getting better.

Learning Positively Correlated with Results

Real learning finally occurred in this stage of the project. Assumptions were challenged. Past errors were corrected and real progress was made.

6.8 EXERCISE – TAP CASE STUDY

1) The bullet points below are the authors' (Van de Ven and Polley) suggestions for improvement. Evaluate each suggestion:
   - Are they actionable? Can a typical person take action to implement the recommendation as stated?
   - Are they likely to be effective? Is the objective clear and will taking the suggested action achieve the objective?

- Separate planning from funding on new ventures to avoid optimistic efforts aimed mainly at securing funding

- Find ways to decrease the "bull index"

- Keep the core team intact

- Keep research and business creation separate to reduce information overload

- Alter agendas and formats of meetings to obtain more candid information transfer

2) Why did the researchers sugar-coat their results?

3) Why did the resource managers publicly accept their story while privately discounting it?

4) Why did researchers ignore poor experimental results throughout most of the project without altering their course?

5) Why did real learning begin in the final period (contraction)? Why didn't it begin earlier?

6.9 TAP CASE STUDY EXERCISE RESULTS

1) Evaluation of the author's suggestions:

Suggestion	Is it actionable?	Will it be effective?
Separate planning from funding to avoid optimistic estimates.	? Not abundantly clear what this means.	Hard to picture what this would accomplish. There is necessary overlap between these functions.
Find ways to decrease the "bull index"	No action is suggested.	
Keep the core team intact	Yes	This should be a goal on all projects.
Keep research and business creation separate to reduce information overload	?	The research is the engine for business creation. The opposite suggestion is frequently made; that researchers need to be more in touch with customers.
Alter agendas and formats of meetings to obtain more candid information transfer	Agendas and formats can be altered, but no concrete suggestion is made of how to alter them.	Assumes that the "sugar coating" was caused by, and can be cured by, changing the meeting agendas and formats. But changing meeting formats is unlikely to get people to say things they don't want to say.

The table above gives an evaluation of the actionability and effectiveness of the Van de Ven and Polley's suggestions. Other than "Keep the core team intact" it does not appear that any of their suggestions are both actionable and likely to be effective.

I argued in Chapter 5 that most published business advice is bad advice. These suggestions certainly reinforce that view. The suggestions of the authors are naïve and are unlikely to solve the identified problems.

2) Why did the researchers sugar-coat their results during the gestation period?

This is not hard to picture. The researchers needed a good story to get funding. In the first two stages the researchers struggled. They realized they had problems. They felt uncertainty and some fear of failure. But their presentations to the resource managers were highly choreographed and optimistic. Were the researchers lying? Possible, but we don't have to make that assumption to explain their behavior. These are probably competent, confident men and women who believed in their process, and believed that they would solve the technical problems. Early failures were just part of the development process. Given enough resources and time, they would get it right if management did not interfere. Sugar coating the results was necessary for them to maintain control of the research effort (avoid management interference).

3) Why did the resource managers publicly accept their story while privately discounting it?

The resource managers probably believed that the researchers would succeed in the end. They likely were willing to overlook fluff because the alternative, challenging the accuracy of the researcher's claims would have been difficult to do. Surfacing negative information and creating negative feelings is avoided when possible. By holding the researchers publicly accountable for achieving their projections, the resource managers sought to maintain control. The resource managers could not surface their concerns to the partner's executives without losing control themselves (the executives would likely have intervened).

4) Why did researchers ignore poor experimental results throughout most of the project without altering their course?

Once the pattern of deception was started it would have been difficult to break. They were probably well aware that some things were going poorly. They were probably working hard to fix the problems before they had to admit them.

5) Why did real learning begin in the final period (contraction)? Why didn't it begin earlier?

When the failures became public it was no longer possible to believe or to pretend that all was well. Outsiders, not married to previous assumptions, were able to challenge those assumptions. Double loop learning was forced upon the team by outsiders.

Summary

The behavior of all participants in this story is consistent with Model I behavior.

6.10 SINGLE VS. DOUBLE LOOP LEARNING – TAP CASE STUDY

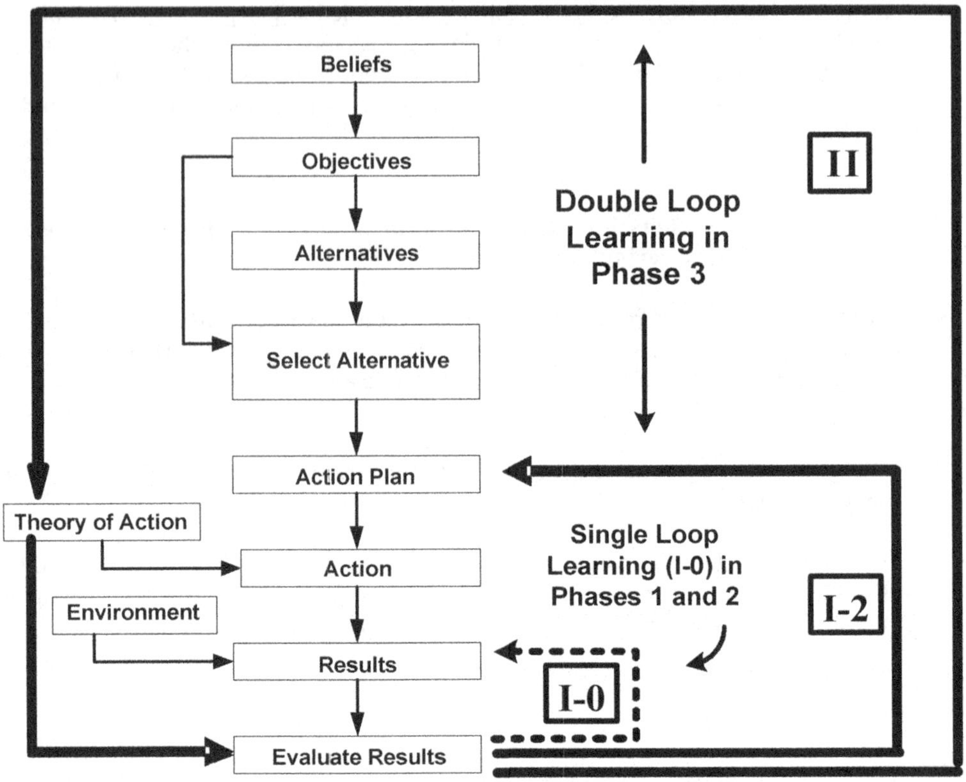

Learning during phases 1 and 2, when it occurred at all, was single loop learning. Learning appears to have been pretty much limited to the I-0 (judging failed results to be successful or anticipating success in the future without significant changes in actions). Model I behavior seriously inhibited introspection into the real results and acknowledgment of real errors.

In phase 3 the researchers were forced to face up to their errors. Visible errors forced focus onto the real causes of error. Researchers modified their evaluation of results and identified real errors in their plans (I-2 level learning) which they began to correct.

Reflective Practice

It is also instructive to consider this story from the point of view of reflective practice (section 5.26). That the researchers had setbacks and ambiguous results is not surprising. They were having a conversation with the environment and the environment was talking back. In the context of typical professional practice it is understandable that they would conceal these results from the resource managers. Shielded as they were from the real results, the resource managers could only be frustrated spectators.

Consider how this project might have proceeded had the researchers really involved the resource managers in their conversation with the situation, openly sharing failed experiments as well as successes and seeking their input. Would the resources managers have lost respect for the researchers? Probably just the opposite would have happened. Had they developed a trusting relationship from the beginning it is entirely likely that the resource managers would have provide useful guidance earlier in the project and they might have 'gone to bat' for the researchers and saved the project when it ran into trouble in the contraction phase.

6.11 GRAND TETON DAM DISASTER

This story is from *Normal Accidents*, by Charles Perrow (6). In eastern Idaho, there is a sparsely populated, farmed section of the Snake River. In 1963 the US Bureau of Reclamation proposed damming the river for flood control and recreation. The proposal was approved by Congress in 1964 and construction began in 1972. There was every reason to expect success of this endeavor. Dam collapses are very rare and no major US government-built dam had ever collapsed. But there was trouble ahead for this dam.

1963 -	**Bureau of Reclamation proposes a dam on the Snake River for flood control and recreation**
1964 -	**Congressional approval**
1972 -	**Construction begins**
Dec, 72 -	**Geologists with the Geological Survey find seismic activity. They hastily write a warning letter.**
June, 73 -	**Warning letter is delivered to Bureau of Reclamation after multiple rewrites, with careful, diplomatic wording which removed all sense of urgency**
1970 -	**Geologist with the Bureau of Reclamation discovers cracks in the rocks. Initially this is a concern. But, the largest crack found is 2". A grouting spec is written.**
1973 -	**With the dam half built, larger cracks are found; some tall enough for a person to stand inside. These cracks are also grouted using the same spec.**
Nov, 75 -	**Bureau of Reclamation began filling the dam at the design fill rate (1 ft/day)**
Mar, 76 -	**Fill rate doubled to 2 ft/day to capture runoff from heavy snowfall for recreation and power reasons. Engineers also note "opportunity to test grouting."**
Apr, 76 -	**Report of groundwater monitoring is issued. Groundwater flowrates are 1000 times the predicted rate. Some of the data is six months old.**
May, 76 -	**Fill rate is doubled again at the request of local congressman**
June 3,4 -	**Three leaks are located. Project Engineer is not worried because leaks are running clear.**
June 5 -	**Two more leaks. One serious.** **Dam collapsed. Project Engineer is worried now!**
Later -	**Project Manager asked if anything could have been done to prevent the accident responded "Nothing"**

The trouble began in December of 1972. Geologists with the U. S. Geological Survey (USGS) were working in the area and discovered that the area was seismically active (potential for earthquakes). They hastily wrote a letter to inform managers at the USGS and the Bureau of Reclamation of the danger. In their letter they noted recent earthquakes in the area, including two of significant strength. They further noted that earthquakes have been known to crack dams and that the weight of a manmade lake could increase the potential for earthquakes. They noted that failure of the dam would cause an enormous flood and concluded, sarcastically, that if the dam were built, that a series of motion picture cameras should be installed to document the collapse.

The letter was considered “too emotional” by their managers. They were forced to redraft the letter several times. It was six months before the “urgent” letter was delivered. By then, the recommendations were as bland as could be contrived with wording such as “We believe that the geological and seismic observations, though preliminary, bear on the geological setting of the Teton Basin Project.” The USGS surveyors didn’t even bother to report subsequent findings because the process proved so difficult.

The warning letter was largely ignored. A Bureau of Reclamation geologist did suggest that the bureau should prepare some “constructive criticism” of the memo.

Before the earthquake letter was received, a Bureau of Reclamation geologist had discovered a simpler problem; cracks in the rock walls. Cracks are a serious problem for earth filled dams. These dams have a concrete shell, but the interior is filled with dirt. The relatively thin concrete shell resists erosion, but is not strong enough to withstand the water pressure on one side. Cracks which allow water to leach out the dirt endanger the dam. Fortunately the cracks were small; the largest was less that 2” wide. A grouting spec was written and the cracks were grouted.

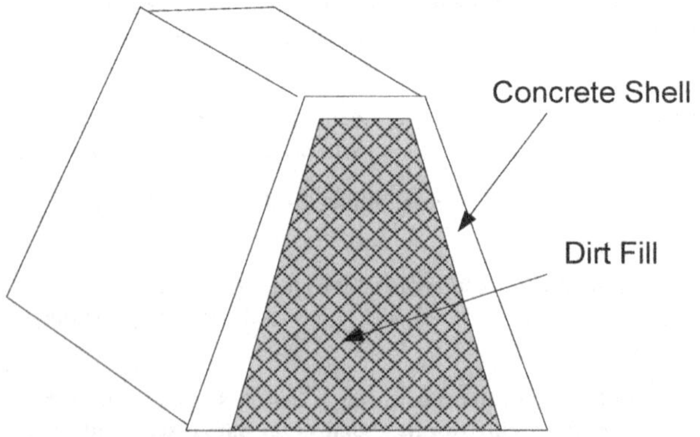

But, in 1973, when the dam was half built, larger cracks, some large enough for a man to walk through were found. The same grouting spec designed for the 2” cracks was used to fill these caves.

The dam was completed in late 1975 and the Bureau began filling it in November, 1975. Filling began at the standard fill rate of 1 foot/day as set in the original design.

Five months later, in March 1976, the project engineer asked for permission to double the filling rate. He noted high snowfalls the past winter and high spring runoff. He noted that this would provide earlier power generation, recreational opportunities, and “an opportunity to test the grouting." They would continue daily inspection for leaks, of course, and "continue" to monitor groundwater for problems.

A month later, a report on groundwater monitoring showed major abnormalities. Several monitoring stations were not working. Some of the ones that were working showed groundwater flowrates over 1000 times higher than expected. This report was ignored. Filling of the dam continued. Indeed, the fill rate was doubled again to 4 feet/day at the request of a local congressman.

Two months later, on June 3rd, two leaks were located and a third was found on June 4th. But they were running clear (indicating no leaching of the dirt fill). The Project Engineer reported that such leaks are common on earth-filled dams and said that he was not worried.

On June 5th two more leaks were discovered, one serious. At 11:57 am, June 5, 1976, the dam collapsed. Fortunately there was adequate warning to evacuate almost everyone below the dam. 11 people were killed. Whole towns were destroyed along with 100,000 acres of farmland and 16,000 head of livestock.

Shortly after the disaster, the bureau issued a publicity release in which they asked themselves "With the benefit of hindsight, was there anything that the Bureau of Reclamation could have done to prevent this disaster?" and their answer to their own question was "Nothing!"

6.12 EXERCISE – GRAND TETON DAM

Identify opportunities for learning in this story.

At these points what actions were taken?

Were these actions consistent with Model I or Model II behavior?

Identify other reasons for errors and failures to learn.

6.13 GRAND TETON DAM EXERCISE ANSWERS

Geological Crew finds Seismic Activity

Action:	Wrote letter
Model:	Hard to judge since we don't have the letter. May have been Model II (attempt to communicate valid and accurate data)

Action:	Supervisor insists on rewrite
Model:	Model I. Diplomatic language, avoid upsetting others, face saving

Action:	Bureau of Reclamation receives letter, but ignores it
Model:	Model I. Stay in control. Admitting error at this stage would have been highly embarrassing.
Other Reasons:	Inertia, anchoring – project well advanced, hard to turn back. False confidence, availability– no major US dam had ever collapsed

2" Cracks found in rock

Action:	Grout spec developed
Model:	Hard to judge. Was grouting a technically adequate solution or was moving the dam to a more suitable location the right answer?

Man-sized cracks found in rock

Action:	Filled using same grouting spec designed for 2" cracks
Model:	Model 1. Cannot face the fact that this might be a bad spot for a dam.
Other Reasons:	Inertia, anchoring. Dam is half built, too late to turn back now. False confidence, availability

Groundwater report – 1000 times predicted rates

Action:	Ignored, no action
Model:	Model 1. Too embarrassing to admit publicly that something is wrong.
Other Reasons:	Inertia, anchoring, false confidence

Dam Collapse – Postmortem question: What could have been done?

Action:	Project Manager answers "Nothing!"
Model:	Model 1. Try to be in control, avoid inquiry, save face, cannot admit errors.

Decision-Making Considerations

Several issues we've covered in the course have bearing on the decisions and mistakes made here.

<u>The dam location decision was made without adequate data.</u> Seismic activity and cracks in the rock walls of the canyon suggested that this was not an ideal spot for a dam.
<u>Anchoring on early decisions:</u> After beginning of construction it was difficult to reassess initial assumptions which proved to be wrong.
<u>Structural secrecy and organizational effects:</u> High level decisions were made by Congress. Congress could not know the problems being tackled by engineers late in the game.
<u>Face-saving, maintaining control</u>: Engineers in charge could not have reported problems found late in the game without serious embarrassment.

6.14 GRAND TETON DAM – HRO PERSPECTIVE

In this section we'll take a look at the Grand Teton Dam disaster from a different perspective.

In the 1980's a team of researchers at the University of California at Berkley conducted a study of organizations that operated hazardous technology yet seemed to have few accidents. They termed these organizations Highly Resilient Organizations (HROs). Weick and Sutcliff (35), using the results of that research, identified five features that make HROs different:

1) Preoccupation with failure rather than success

   It is easy to become complacent because serious accidents are rare. HROs guard against complacency by focusing on potential failure and always being on the lookout for cues that something is going wrong.

2) Reluctance to Simplify

   Cues that something is going wrong can be hard to distinguish from "noise". HROs are mindful of this and do not readily dismiss cues of potential problems. Everyone is encouraged to identify and report potential problems.

3) Sensitivity to Operations

   Managers at HROs keep abreast of what is happening "on the ground."

4) Commitment to Resilience

   HROs plan for the unexpected. Minor problems don't get out of control because there are plans in place to deal with them. Also, people on the front line know that they have the authority to do whatever is necessary to deal with an urgent problem.

5) Deference to expertise rather than bureaucracy

   Emergencies must be managed by the people with the expertise to understand and deal with the problem. Management must yield to the experts during an emergency.

How does the Teton Dam design team measure up against these criteria?

The Teton Dam team seemed willing and eager to simplify problems. The caulking issue in particular comes to mind. Caulk specified for 2" cracks was used to fill "caves".

Deference was clearly to authority rather than to expertise in the response to the USGC geologists warning about earthquakes. It may also have played a role in increasing the fill rate at a Congressman's request.

They seemed distant from operations as judged by the fact that they ignored the groundwater flowrate information.

Most of all they were preoccupied with success rather than failure and ignored multiple cues that something was wrong.

6.15 LEARNING FROM COMPUTER PROGRAMMERS

We have many opportunities to learn from others doing similar work. Many of the problems we face in engineering design are also faced by computer programmers. An interesting design process applied in that field is the process called Extreme Programming (XP). In this section we'll discuss what we might learn about engineering practice from XP.

Extreme Programming – Summary

For an excellent description of XP see *Extreme Programming Installed* (26). XP is summarized below.

- **Begin with user stories or use cases**
- **Select user stories with the highest business value. Work on these first.**
- **Test everything continuously. Write the user story tests before working on the code. Make sure everything works all the time.**
- **Integrate early. Get everything working together while it's still easy to do.**
- **Release working code early and often. Deliver value early. Provide opportunities for customer feedback.**
- **Keep it simple. Write the simplest code that will achieve the objectives.**
- **Do it only once. Make changes easy by doing discrete work only once.**
- **Work only on the current user story. Don't pre-invest in infrastructure you may need later. (Because you may in fact not need it.)**
- **Joint ownership of code. Everyone owns all the code and is empowered to fix errors.**
- **Pair programming. All programming is done in pairs; one person programs, the other watches and consults.**

User Stories or Use Cases

XP begins with the writing of user stories or use cases. A user story is a simple description of a customer objective. Stories are kept simple such that a story can be programmed in 1 to 4 weeks. This yields manageable chunks of work useful for project estimating and tracking. It also provides a mechanism for prioritizing work.

Use cases are more formal and detailed (see section 3.6).

Work Highest Priority User Stories First

Dividing the project into user stories allows for prioritization. Work first on the user stories with the highest business value.

Test from the Beginning – Continue Testing until the End

This is the most interesting feature of XP to me. Every user story is tested as it is built. In fact, the tests are written first; <u>before</u> the code. The objective of the user story programming is then to pass the user story test. A story is finished when the unit test for that story is passed.

The tests are all automated and are run frequently – often several times a day. Any programming which causes previously successful code to fail is identified immediately and fixed. All code works all the time!

Integrate Early

Integration of complex systems is difficult. Saving integration until the end causes schedule delays and cost overruns as unexpected interface problems are discovered. XP stresses integration early while the systems are simple, even if some systems are just data generating dummy systems to test communication.

Release Code Early and Often

Opportunities are eagerly sought to release working code that satisfies some of the customer's requirements early. This provides value to the customer early, generates success experiences which build moral and provides opportunities for feedback from the customer.

Incrementally Change Scope Based on Feedback

Feedback is used to adjust objectives and priorities. As the customer sees what is developed her vision of what is possible, useful and necessary is likely to change. XP is flexible enough to adopt these changes – you simple change the impacted user stories.

XP does not make use of overall system specifications up front because these are difficult to generate and rarely accomplish what you hoped for anyway.

Keep it Simple, Work only on the Current Story, Do it only Once

Complicated code is viewed as a source of future bugs and maintenance problems. XPers write the simplest code that will work.

XPers work only on the current user story. No effort is made to write general purpose applications (not simple) that have more power than what is needed to solve the current user story. No effort is put into developing infrastructure not currently needed, even if it is likely that such infrastructure will be needed later. Diverting efforts from the current work is avoided at all costs.

XPers do things only once. For instance, data needed from a database is read in only once so that if the database structure is changed, required changes to the program are minimized.

Joint Code Ownership

In XP everyone owns all the code. Anyone who finds a mistake in the code is empowered and expected to fix it. The unit tests ensure that the changes are successful and haven't broken anything else.

Pair Programming

An interesting feature of XP is that all programming is done by a pair of programmers. One programmer is the driver – she types in the code. The other, the observer, continuously watches the work of the driver actively watching for errors and discussing what's happening and should happen. With time to think, the observer catches both trivial errors (syntax, spelling) and strategic errors. The two programmers, both completely aware of what is happening can brainstorm solutions to tricky problems. XPers claim that pair programming is more accurate and more efficient and improves moral.

What Can we Learn from XP?

There are several things engineers might learn from XP:

- Use Cases are described in Chapter 3. They are very useful for identifying objectives and insuring completeness and clarity in specifications. Use Case extensions are useful for identifying things that might go wrong.
- Releasing code early and often is a very useful insight. Engineering is an iterative process. It is our tendency to do all the iterations in isolation in order to deliver a finished product to the client/customer. Sharing intermediate work can be very useful. When this is done the client feels included, gets a feel for the tradeoffs involved, get to refine his objectives and is not surprised at the end of the study. (See discussion of reflective practice in section 5.26).
- Pair programming does not seem immediately useful for design efforts, but the idea is in fact used in hazardous industries where two operators or two mechanics are used to do the work of one. The rational is the same; two heads are better than one; one person has time to look for hazards while the other does the work.

I want to focus on the continuous testing idea.

<u>Continuous Testing</u>

It is much more difficult to automate testing in engineering work than in computer programming. But thinking about this has led me to some insights. The first is that much of what we call design is, in fact, largely analysis or testing. For example, pipe comes in standard diameters, wall thicknesses and materials of construction. When we specify a pipe we select a standard product and then <u>test</u> it against our objectives which may include velocity limits, pressure drop limit, pressure rating requirements, corrosion resistance, etc. If the selected pipe passes all the tests, we use it. Such tests could be automated and run each time the material balance changes.

For analytical work routine testing is probably doable; how about synthesis. Much of engineering work involves the synthesis of solutions. Often design changes have unanticipated consequences. Design changes are typically driven by the need to fix a particular problem. The design team is often in tunnel vision mode when fixing problems, especially late in the project. The identified problem is fixed without giving attention to the rest of the project objectives. *The learning to take away here is not that the tests need to be run frequently as in XP, but that the tests (objectives) should be made explicit so that tests can be run when needed.*

6.16 ORGANIZATIONAL LEARNING

So far our discussion has tacitly assumed that the learning is being done by an individual. Since engineering is a team sport, we need to be at least as interested in organizational learning as we are in individual learning.

What is Organizational Learning?

As individuals we learn by processing information via our central nervous system. Organizations do not possess a central nervous system. In order for organizations to learn they must possess analogous structures that enable their members (who can learn) to collect, analyze and apply information as a group. Organizations learn when information is processed by groups of individuals within the organization resulting in shared mental models and/or revised processes or procedures. Where individual learning is a mental process, organizational learning is a social process.

Organizations capture and disseminate knowledge for future use via Organizational Learning Mechanisms (OLMs). OLMs take many forms including lessons learned studies, mentoring, classroom training, writing procedure and standards, etc. The effectiveness of an organizations OLMs together with the effectiveness of the organizational structure largely determine how effective the organization will be in bringing individual knowledge to bear on organizational problems.

We can argue that an organization knows something if
1) at least one individual member knows it and
2) the organizational structure, processes and culture make it likely that the knowledge will be effectively brought to bear on relevant problems faced by the organization.

Definition:

Organizational learning is an increase in the knowledge or skills of individual members of the organization and/or a change in the structure, processes or culture of the organization which enables the organization to be more effective at planning and implementing actions which achieve the organization's objectives.

Classifying OLMs

Lipshitz, Friedman, Popper (34) have defined a two-dimensional scheme for classifying OLMs derived from the two basic questions:

"When does the learning occur relative to when the task is performed?": Online vs. Offline OLMs

Online learning occurs when working and learning occur simultaneously. This occurs when we experiment on-the-job with a different way of doing a task and note whether the experiment was successful or not. Classroom learning, on the other hand, is usually not immediately applied.

"Who carries out the learning?": Internal vs. External OLMs

Learning is Internal when the actor is also the learner. Learning is External when a teacher instructs those who will perform the task.

The table below shows the classification of commonly encountered OLMs.

	INTERNAL	EXTERNAL
OFFLINE	*Lessons learned* *After-action reviews* *Communities of practice* *Self-study* *Literature review* *Writing paper for conference* *Developing process for personal use*	*Scenario planning* *Seminars* *College courses* *Conferences* *Project audits*
ONLINE	*On-the-job Training* *Online experimentation* *Online debriefing*	*Coaching, Mentoring* *Peer reviews* *Codes and Standards* *Work Processes / Procedures* *Training manuals* *Checklists*

While it may not be important which particular OLMs an organization chooses to use, I would expect an effective learning organization to make use of one or more OLMs from each classification.

6.17 LEARNING THE OBVIOUS AND MISSING THE POINT

Before leaving the subject of learning I want to discuss briefly a class of learning errors we can label as systems errors. It is tempting, given a particular history, to craft simple explanations for events. We labeled that behavior sensemaking back in Chapter 2. We often learn the wrong lessons because we focus on the obvious causes of events and miss the systems interactions. The following story from *Learning from Accidents* by Trevor Kletz illustrates the point.

A leak in a small bore line caused by a poorly made-up joint led to an explosion that killed four people, injured several others and utterly destroyed a compressor building.

The natural reaction to this story is to blame the craftsman who made-up the joint. There are two good reasons why you should not make that assumption:

1) Model I behavior will make it hard for you to investigate any accident if your focus is on placing blame.

2) Most accidents have multiple causes. Focusing on the obvious "cause" will not help you avoid recurrence if other factors that contributed are not identified.

A more thorough investigation, directed by the question "How can we prevent this from happening again?" yielded valuable insights including:

Management Failings

- There were no established guidelines for joint integrity.
- Management had recently changed the rules for who could make up high tech joints. Previously only trained and tested mechanics could work on them. The change was made to improve maintenance efficiency.
- Leaks were frequent at this plant, but they were tolerated because "there were no ignition sources" in the plant.

Engineering Error

- The company had a policy of putting compressors outdoors because of the known hazard of contained gas clouds. Never-the-less, this compressor was located in a building because maintenance of an outdoor compressor is difficult in cold weather.
- There were no gas detectors in the area.

Under-reaction by Operators

- The operators knew about the leak for several minutes before the explosion. They were in the process of shutting down the plant. But they took no action to clear the area. They believed that there was no danger of explosion because they believed there were no ignition sources in the area. The dead and injured were mainly mechanics working on a nearby compressor who had no need to be in the area.

The obvious cause of this accident was a poor joint installation. An investigation seizing upon the obvious cause would have missed many important points.

6.18 CHAPTER SUMMARY – FACTORS INHIBITING LEARNING

Learning is impeded by Model I behavior. We have a need for control, a need to win, difficulty in believing we could be wrong, difficulty in reevaluating assumptions, a wish to avoid embarrassment and to avoid embarrassing others.

We tend to learn the least threatening lessons from an event. Whenever the results are ambiguous we will:

- tend to judge the results a success if possible,
- tend to blame failure on bad luck rather than error,
- tend to blame slips, lapses or rule-based errors rather than knowledge-based errors (flawed plan),
- tend to blame a flawed plan rather than judgment errors, and
- tend to blame judgment errors rather than flawed beliefs.

Learning is complicated by Systems Effects. There are usually multiple causes of an accident or error. Isolating an obvious cause is often not the whole story and can lead to erroneous learnings.

As individuals we learn by processing information via our central nervous system. Organizations do not possess a central nervous system. In order for organizations to learn they must possess analogous structures that enable their members (who can learn) to collect, analyze and apply information as a group. Organizations learn when information is processed by groups of individuals within the organization resulting in shared mental models and/or revised processes or procedures. Where individual learning is a mental process, organizational learning is a social process.

Organizations capture and disseminate knowledge for future use via Organizational Learning Mechanisms (OLMs). OLMs take many forms including lessons learned studies, mentoring, classroom training, and writing procedures and standards, etc. The effectiveness of an organizations OLMs together with the effectiveness of the organizational structure largely determine how effective the organization will be in bringing individual knowledge to bear on organizational problems.

We can argue that an organization knows something if

1) at least one individual member knows it and
2) the organizational structure, processes and culture make it likely that the knowledge will be effectively brought to bear on relevant problems faced by the organization.

Organizational learning is an increase in the knowledge or skills of individual members of the organization and/or a change in the structure, processes or culture of the organization which enables the organization to be more effective at planning and implementing actions which achieve the organization's objectives.

6.19 ANSWERING THE FOUR QUESTIONS

What observations from this chapter can we make about our four key engineering questions?

Why do so many changes occur late in projects?

Learning/admitting that there is an error in your design is difficult.

Learning is often delayed until the error surfaces in some public way that cannot be denied, as when someone not involved in the design flags the problem.

Why do we have so many problems at interfaces?

Learning that others are wrong

When there is conflict it is relatively easy to see error in the other team's approach and difficult to see errors in your own or your team's work.

Why do we repeat past mistakes?

Simply recognizing a problem is not learning:

While recognizing that you have a problem may lead to learning, it is not learning. Learning comes from actually correcting the problem. We don't work hard enough on problem solutions because of the invalid view that once we've identified the problem we've got it solved.

Failure to learn difficult lessons (double-loop learning) due to Model I behavior:

We are expert at learning simple, non-threatening lessons. We have difficulty learning anything that is threatening or potentially embarrassing.

Lack of Motivation to Change:

Real learning occurs when we take action to solve problems which are important to us. We may continue in a failed course of action if the results are not that important to us.

Accidents often have multiple causes:

Most industrial accidents result from multiple things going wrong. The obvious cause of a problem may not be the only cause or even the most important.

Sensemaking:

The history of past projects is not unambiguous. We never know all the factors that went into making a project successful or unsuccessful. It is easy, given a partial history, to craft an explanation for a past error that is wrong or incomplete.

Individual Learnings not utilized by the Organization:

Though the introduction to organizational learning presented in this chapter was very brief it is easy to see that lessons learned by one or a few persons may not be available to the organization as a whole.

Why do we so often finish behind schedule?

Failure to learn from missed schedules:

Since we routinely miss schedules there are multiple opportunities for learning the reasons why, yet we do not. This is a particular instance of failure to learn from past mistakes.

<u>Learning the wrong lessons from missed schedules (sensemaking):</u>

The history of past projects is not unambiguous. We never know all the factors that went into making a project successful or unsuccessful. It is easy, given a partial history, to craft an explanation for a past schedule failure that is wrong or incomplete. It is also easy to attribute failure to bad luck.

Chapter 7
Naturalistic Decision Making
(Impacts of stress and expertise on decision making)

Sources:

This chapter is based largely on the work of Gary Klein as reported in his book *Sources of Power* (3).

Additional information is obtained from *Decision Making under Stress* edited by Flin, Salas, Strub and Martin (10).

Significant insights on pattern recognition come from the work of James Reason on human error (28).

Preface:

The information presented in the first six chapters of this course is based on two tacit assumptions:

1) The decision maker is not under any time pressure or stress.
2) The decision maker does not have any particular expertise in the subject matter.

But most important decisions are made under some stress or time pressure and are made by persons with some expertise in the subject. In this chapter we investigate the impacts of stress and expertise.

Pressure is what you feel
when you don't know what you are doing.

Chuck Knoll and others

CHAPTER 7 CONTENTS

7.1 NDM INTRODUCTION
7.2 NATURE OF STRESS
7.3 IMPACT OF STRESS AND EXPERTISE – PRELUDE
7.4 STORY: THE OVERPASS RESCUE
7.5 STORY – TANK FARM FIRE
7.6 RECOGNITION PRIMED DECISION MODEL
7.7 FOCUS ON SITUATION ASSESSMENT (rather than moment of decision)
7.8 SOURCES OF STRESS
7.9 MINIMIZING STRESS
7.10 SOURCES OF POWER
7.11 POWER OF PATTERN RECOGNITION
7.12 POWER OF INTUITION
7.13 THE POWER OF MENTAL SIMULATION
7.14 COGNITIVE TASK ANALYSIS (CTA)
7.15 TEAM CTA
7.16 CTA IN LESSONS-LEARNED STUDIES
7.17 IMPROVING PERFORMANCE IN STRESS-FILLED JOBS
7.18 RULES AND PROCEDURES
7.19 CHAPTER SUMMARY
7.20 ANSWERING THE FOUR QUESTIONS

7.1 NDM INTRODUCTION

> What happens when real people try to make decisions
> in the real world,
> under time pressure or stress,
> with limited data,
> about things they care about and
> about which they have some expertise?

The data presented in the first six chapters of this course was developed by researchers in many universities and companies all over the world. Much of this research was done in laboratory settings, often with students as subjects. In order to minimize bias in these studies, the subject matter is usually something generic. The research subjects usually know little about the subject matter, or all know about the same amount (no differences in expertise). And the subjects are usually not under any significant stress.

But many of the important decision we make are made under stress; and in most cases we have some expertise in the subject. It was not until the mid 1980's that anyone began to seriously investigate the impact of stress and expertise on decision making. Since that time a very vibrant field of study called Naturalistic Decision Making (NMD) has evolved. NDM is

> **"The study of how experienced people, working as individuals or groups in dynamic, uncertain, and often fast-paced environments, identify and assess their situation, make decisions, and take actions whose consequences are meaningful to them and to the larger organization in which they operate."** Zsambok in *Decision Making Under Stress* (10)

Or more simply:

> **NDM is the way people use their experience to make decisions in field settings."** Zsambok and Klein in *Decision Making Under Stress* (10)

Stress impacts our decision making. Learnings from laboratory studies of subjects not under stress do not translate well to real world situations.

Expertise impacts our decision making. Learnings from laboratory studies of subjects with no subject matter expertise do not translate well to the real world.

7.2 NATURE OF STRESS

Definition:
1) Stress is an environmental demand on social, biological or physiological systems. We experience stress when the demand exceeds our ability to easily cope.
2) Stress is a perceived requirement to act without sufficient data, knowledge, time or resources to effectively plan and execute the required actions.

The stress we feel is a function of the imbalance between the percieved demand and our percieved ability to cope with the demand.

Effects/Symptoms - Imbalance between Demand and Ability:

While we can define stress as a demand or requirement to act, we tend to define it in terms of its symptoms. Symptoms can be physical, mental and biological. Some common ones are:

Physical	Mental	Behavioral
Increased heart rate	Confusion	Talking rapidly
Sweating	Worry	Under communicating
Butterflies in stomach	Overwhelmed	Excluding others
Rapic breathing	Feeling ill	
Tense muscles		
Dry mouth		

7.3 IMPACT OF STRESS AND EXPERTISE – PRELUDE

Reactions to Stress

Stress impacts multiple areas of our decision-making process. Stress has important impacts on Data Collection and Evaluation including:

Acceleration:
: Under time pressure we speed up our data collection and data evaluation efforts.

Filtering:
: When we don't have time to collect much data we filter or select the data that we pay attention to.

Tunnel Vision:
: Once we see a likely candidate as the cause of the problem we focus on it to the exclusion of others.

The second step in the decision-making process, Identifying Objectives and Alternatives is also impacted significantly:

Objectives Identification:
: Under stress we tend to identify only the most obvious objectives. We don't have time to thoroughly evaluate the situation and comprehensively identify all the valid objectives. To an extent, our objectives will be limited to what we think we can possibly achieve.

Sequential vs. Simultaneous Search:
: Under time pressure we do not try to identify a large number of possible solutions. We tend to focus on the first idea that comes to us and use it if we think it will work, moving on to another option only if we find major fault with the first one.

The deciding and action steps of the decision process are also impacted by stress. Our goal in making the decision changes:

Satisfying vs. Optimizing:
: Under pressure we no longer seek the best or optimal answer, we seek any solution that will work.

Stress also takes a toll on us physically and mentally. We sweat, worry, and get "butterflies" in our stomachs. We get tense, irritable, anti-social. We get confused, feel overwhelmed. We may tend to make emotional reactions (such as denial or distancing) and/or make emotional decisions. Perhaps worst of all, under conditions of stress we are less likely to communicate effectively.

But

Some level of stress improves decision making by creating a sense of urgency.

7.4 STORY: THE OVERPASS RESCUE

This story is from Klein (3)

A fire-fighting company is called out to rescue a young woman who has fallen or jumped from an overpass. Rather than falling to her death, she has landed on the metal supports of a highway sign and is precariously balanced there, semi-conscious.

The lieutenant arrives and surveys the scene.

Two of his men climb out to support the woman.

He sends a hook and ladder truck down to block traffic on the freeway below.

Now the question of how to pull her to safety. The usual method would be a safety harness. But she would have to be in a sitting position for them to be able to attach it. He pictures her falling off the support as they try to get her into a sitting position and abandons that idea.

He next considers attaching the safety harness from behind her (backwards). He envisions lifting her in that position and concludes that the stress on her back would be too great.

Next he considers a rescue belt, but it too must be installed from the sitting position.

Then he has a novel idea. Why not use a ladder belt. They are the wide leather belts that firemen use to tie off to a ladder so that they'll be supported if they lose their step. He thinks that will work. His instructs his men to do it.

Decision Making by the Lieutenant

A striking feature of this story is that the lieutenant never makes any decisions in the sense of decision making we discussed in the first six chapters of this course. We expect him to make a list of alternatives, evaluate them, select the best and implement it. But he never makes a list and never compares one alternative to another. So how does he determine a course of action?

He mentally simulates each idea as it occurs to him. He imagines the results of each, until he finds one that appears to work. He isn't comparing alternatives to each other. Rather he is comparing each alternative to some desired outcome.

We might guess that a novice would approach the problem this way, but that an expert would identify several possible courses of action and carefully deliberate about the merits of each different course.

The next story sheds light on that.

7.5 STORY – TANK FARM FIRE

This story is from Klein (3).

The mid-west is dotted with oil tank farms. Oil is pumped up from oil fields in Texas and Oklahoma and stored in these tank farms. One bitterly cold night one of the tanks bursts open and the oil catches fire. Two nearby tanks soon catch fire. Burning oil is pouring down a ditch blown by the wind.

The fire departments from local small towns, all manned by volunteers, report to the fire. The crews are huddled against the cold. They can barely recognize members of their own crew. It's not clear who is in charge. They are afraid other tanks will catch fire. They don't know whether to evacuate the town or not.

One commander orders a trench dug to contain the burning oil. Another decides to pump the oil out of a tank engulfed in flames, but is not sure if the lines are intact. They try spraying foam on top of the endangered tank, but the wind blows it away.

At this point they stop fire-fighting efforts and regroup to plan. They list options, evaluate the advantages and disadvantages of each. They make a couple more ineffective attempts before giving up and calling in firefighting specialists from Boots and Coots.

The experts from Boots and Coots arrive the next day, take a quick look at the scene and announce that they will need a great deal more foam. The local commanders inform them that they have don't have much foam. The Boots and Coots men reply "Of course not, we've already ordered it. It'll be here tomorrow."

From that point on, under the direction of experts, the fire-fighting efforts go smoothly. Within two days the blaze is out.

Decision Making – Novices vs. Experts

Here we have novices initially fighting a fire replaced later by experts. Note that it is the novices who make a list of alternatives and argue the pros and cons. The experts take a quick look at the fire, formulate a plan and start fighting the fire. And they do so effectively.

Discussion:

How will the behavior of experts differ from the behavior of novices in time pressure situations?

Will experts or novices identify more alternatives for solving the problem?

Will experts or novices feel more stress?

7.6 RECOGNITION PRIMED DECISION MODEL

The classic decision model:

gather data,
identify objectives,
list alternatives,
select the best alternative

does not describe the decision-making process used in these stories.

Observations such as these led Klein (3) to propose an alternative decision model. The graphic below is a simplified schematic of Klein's Recognition Primed Decision Model (RPD). RPD has been very successful at describing how experts make decisions under time pressure.

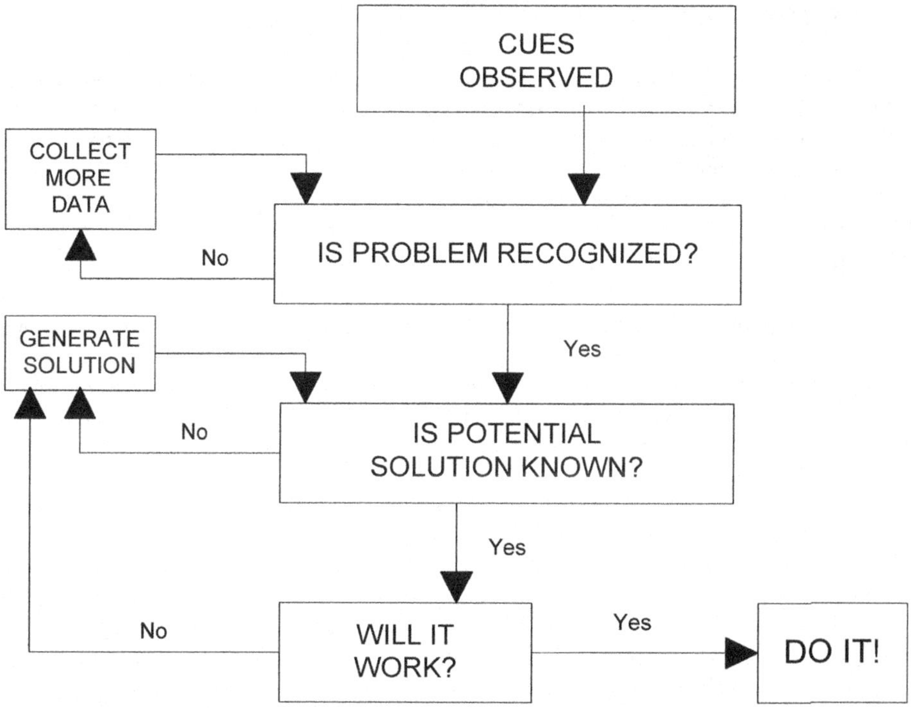

RPD Description

1. The process begins with observed cues that there is a problem.
2. The expert witnessing the cues asks "Do I recognize this problem?"
3. If she does not recognize the problem, she will gather more data in her attempt to diagnose the problem.
4. If she does recognize the problem she next asks herself "Do I know a possible solution?".
5. If the answer is "No", she must do some problem solving to attempt to identify a solution.
6. Once she has identified a potential solution she asks "Will it work?"
7. If she believes the solution will work she implements it.
8. If the answer is "No" she must cycle back to problem solving to identify another possible solution.

Experts vs. Novices

This model is employed by experts. It is their expertise that allows them to recognize a problem quickly from the cues provided and to quickly identify possible solutions. It is also expertise which allows them to evaluate possible solutions and to develop enough confidence in their selected solution to implement it without extensive comparison of alternatives.

Novices do not possess the tools of pattern recognition nor the mental library of possible solutions nor the systems knowledge necessary to evaluate potential solutions "on-the-fly". Hence, novices are forced to be more structured, to collect more data, and to list and evaluate alternatives before making a decision.

Use of RPD without Stress

One strength of the RPD process is speed. Under time pressure we don't have time for analytical decision making. RPD permits experts to make effective decisions under time pressure. It should not be surprising to find that we apply the method routinely even when there is no time pressure.

Strengths of RPD

- It explains how people make decisions without evaluating the pros and cons of multiple alternatives.
- It explains how people use their experience to adopt the first course of action they think of or the first that they think will work.

7.7 FOCUS ON SITUATION ASSESSMENT (rather than moment of decision)

Perhaps the biggest difference between research in analytical decision making and research in NDM is the area of focus. Research in analytical decision making focuses on the moment of decision and hence on methods of evaluating competing alternatives. The key focus of NDM is on situation assessment.

Recognizing the problem is fundamental to applying RPD. Situation assessment depends on pattern recognition, the ability to judge a situation as typical.

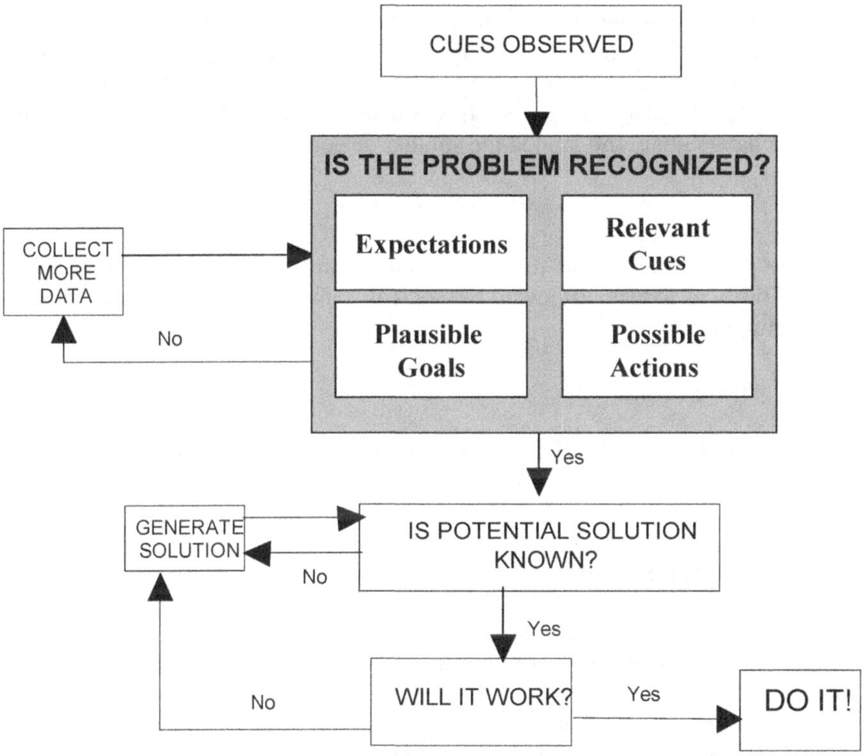

Pattern Recognition by the Automatic Subsystem

We noted in chapter 5 that pattern recognition is one of the fundamental skills of our automatic (unconscious) cognitive subsystem. This makes recognition of learned patterns automatic and very fast.

Pattern recognition has four byproducts:

- Expectations
- Identification of plausible goals
- Identification of relevant cues
- Identification of possible actions

<u>Expectations</u>

When we recognize a situation, we become able to predict future developments. Expectations about future developments are a powerful check on our understanding of the problem. If we correctly predict what will happen next, we gain confidence in our diagnosis of the problem. If our expectations don't come true, we must look for another explanation.

<u>Plausible Goals</u>

When we recognize a situation, we can judge what goals make sense to pursue. A pilot loses an engine. She knows that she can fly safely on his one remaining engine, but does not want to fly any further than necessary. But suppose that the nearest airport is a small town municipal airport with a very short runway. In a dire emergency she would land there, but judges that it is safer to fly a little farther to an airport with a longer runway. Her recognition of the problem and familiarity with the plane and with the surrounding area allows her to identify plausible goals. If she detects any trouble in the remaining engine, she might change goals and land at the smaller airport.

<u>Relevant Cues</u>

Recognition of the problem allows us to focus on cues that would reinforce or disprove our conclusion. Ability to focus on relevant clues greatly improves our efficiency in data collection and evaluation.

<u>Possible Actions</u>

Recognition of the problem should yield ideas on courses of action to solve or mitigate the problem.

Sources of error in pattern recognition:

The story is not all positive. The very power of pattern recognition causes some limits and weaknesses. James Reason (28) identified these:

<u>Bias for strong rules</u> when input data is ambiguous. Strong rules are those which have worked well frequently in the past. (see section 5.11)

<u>Confirmational bias</u>. Initial data is more salient that later data. We tend to form impressions early (anchors) and to use later data to confirm our initial hunches. Evaluation of later data is biased in favor of confirming initial opinions.

<u>Relevance bias.</u> Once we've determined that certain cues are relevant we spend most of our time on those and ignore other cues that might disconfirm our beliefs.

7.8 SOURCES OF STRESS

It is easy to see where in this process people will feel stress. Stress occurs where we cannot meet the demands of the environment.

> *If we immediately recognize the problem, and*
> *know immediately and with confidence how to solve it,*
> *we will feel little, if any, stress.*

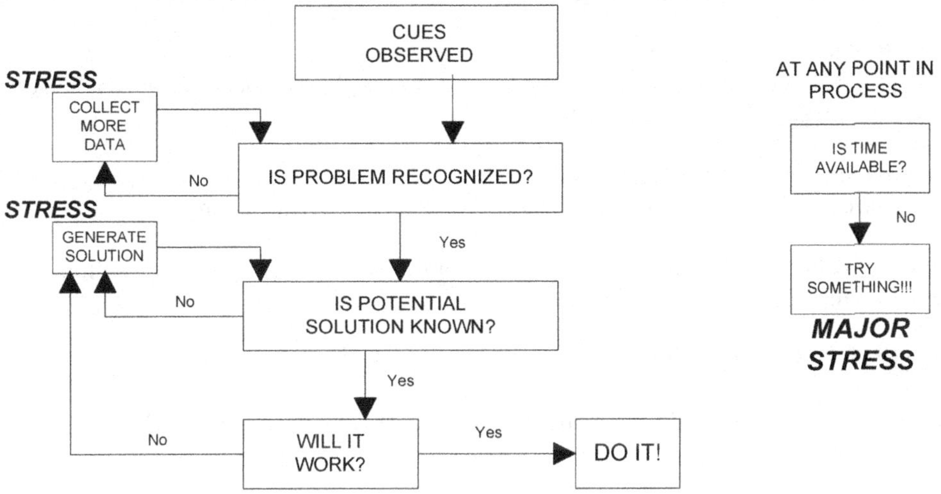

Conversely,

If we do not recognize the problem given the first cues, we have to collect more data. We will feel stress at that point.

If we recognize the problem, but don't know how to solve it, we need to do some quick problem solving. We will feel stress at that point.

If we judge that our ideas for solving the problem won't work, we need to cycle back to problem solving. We will feel stress at that point.

If, at any point, we feel like we are out of time we will feel major stress.

Causes of Stress:

<u>Situation Ambiguity</u>

The less certain you are about what the problem is, the more stress you will feel. Situation ambiguity can result from:

Ambiguous cues (noise, vibration, etc.)
Cues with multiple possible interpretations
Conflicting cues

Solution Ambiguity

The less certain you are about how to solve the problem; the more stress you will feel. Three possible situations exist:

Rule-based solution:
The situation has been anticipated and a rule or procedure exists for dealing with it.

Choice decisions:
The decision maker must choose between multiple available options.

Problem Solving:
No solution is readily available or recognized. The decision maker must be creative in coming up with a solution.

Stress typically increases progressively through this list; lowest stress if a rule exists and highest stress if a solution has to be created via creative problem solving.

A key skill here is mental simulation - the ability to evaluate potential solutions by predicting the future course of events.

Amount of Time Available:

The shorter the time available (or perceived available) to solve a problem, the more stress you will feel.

Level of Risk:

The more there is riding on the decision, the more pressure you will feel.

7.9 MINIMIZING STRESS

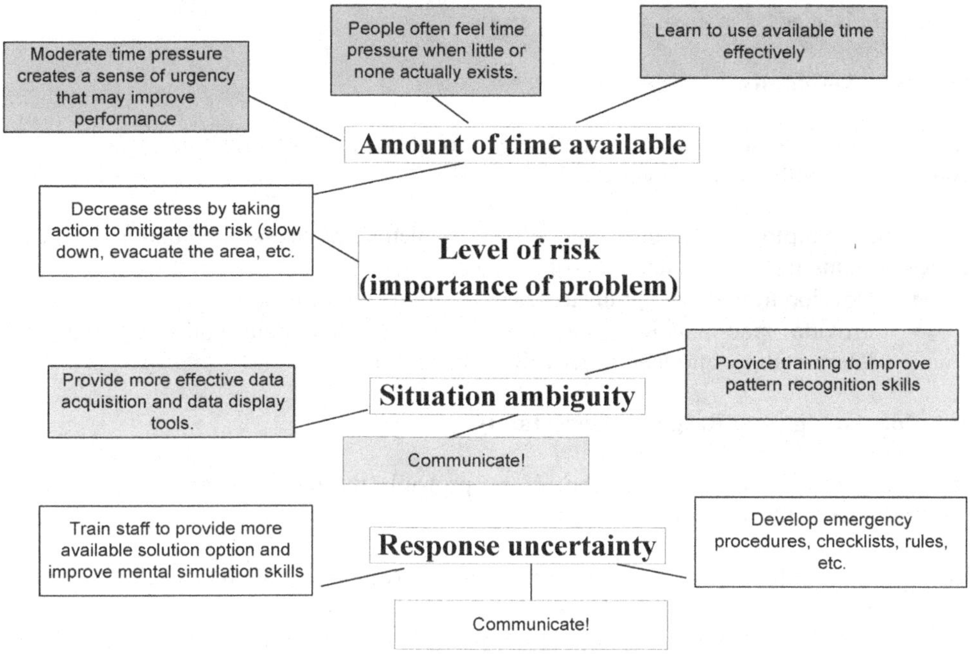

This section discusses ways of minimizing stress.

Amount of Time Available:

You feel stress if you don't have time to solve a problem. In this situation there are several things you can do.

First, recognize that <u>some time pressure can make us more efficient and effective by creating a sense of urgency</u>. Don't let manageable time pressure become a problem.

A related observation is that <u>most people underestimate the time available, and hence rush even when there is time available</u>. Carefully identify how much time you really have. Don't rush unnecessarily.

<u>Try to use the time you have effectively</u>. If you don't understand the situation, collecting more data is a better use of time than worrying. If there are multiple issues concentrate on the most important. People tend to communicate less effectively under stress. Communicating with knowledgeable coworkers might be the best use of time.

If there is an immediate threat, <u>consider how you might "buy time".</u> Actions such as evacuating unnecessary personnel and putting emergency resources on alert might not help you solve the problem, but might give you enough breathing room to solve it without panicking.

Level of Risk:

The greater the risk or potential damage, the more stress we will face in making the decision. When faced with risk consider ways to minimize the impact even if you don't know how to solve the problem. In an emergency, evacuating the area so that only equipment is at risk can greatly decrease stress.

Situation Ambiguity

If you don't understand what's happening, probably the most effective response is to communicate with other(s) that may.

To prepare employees to correctly assess problems in stressful situations develop effective training programs and control systems:

- Develop training programs to teach pattern recognition skills.
- Provide man-machine interfaces which simplify data gathering and data interpretation, and that effectively manage alarms.

Solution Ambiguity / Response Uncertainty

If you don't know how to solve a problem, probably the most effective response is to communicate with someone who might.

To prepare employees to respond correctly in stressful situations develop effective training programs and control systems:

- Develop training programs to teach pattern recognition skills and mental simulation skills.
- Provide training to facilitate pattern recognition skills and mental simulation skills.
- And develop procedures, systems and processes for responding to emergencies.

7.10 SOURCES OF POWER

We are capable of making decisions efficiently in those fields in which we have expertise. The ability of experts to generate good solutions quickly, as described in the RPD model, requires skills of many sorts. These are cataloged in *Sources of Power* (3). Three of them are discussed in the next three sections:

- The power to see patterns (pattern recognition)
- The power of intuition
- The power of mental simulation

7.11 POWER OF PATTERN RECOGNITION

Where novices see raw data, experts see patterns. Experts can size up a situation at a glance and realize that they have seen it, or variants of it, before. This skill enables application of the Recognition Primed Decision Model.

The situation awareness that results from pattern recognition allows experts to focus on relevant data. Hence, experts are less likely to suffer from information overload.

Chapman's Homerun

My favorite example of pattern recognition is the ability of baseball outfielders to judge a fly ball. This is learned pattern recognition even though baseball players cannot describe the pattern.

Seville Chapman (P7) identified the pattern. Watch an outfielder when a ball is hit to him. If the ball is going to land right where he is standing he will not move much; he will wait for the ball. If the ball is hit to land very far away from him he will sprint towards the point where the ball will land, perhaps arriving there just in time, perhaps too late. There is nothing noteworthy in this story so far.

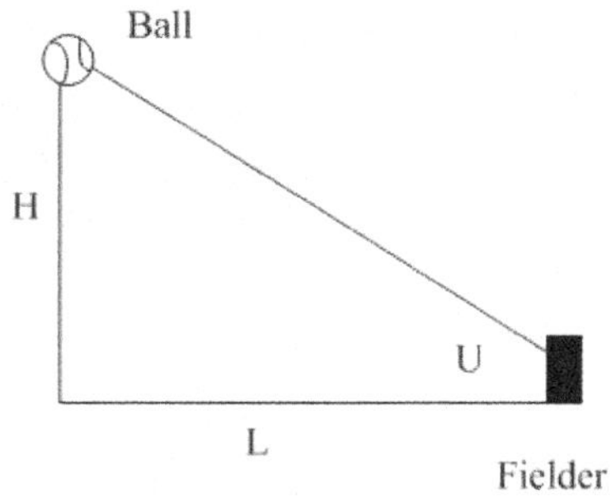

Tan U = H/L

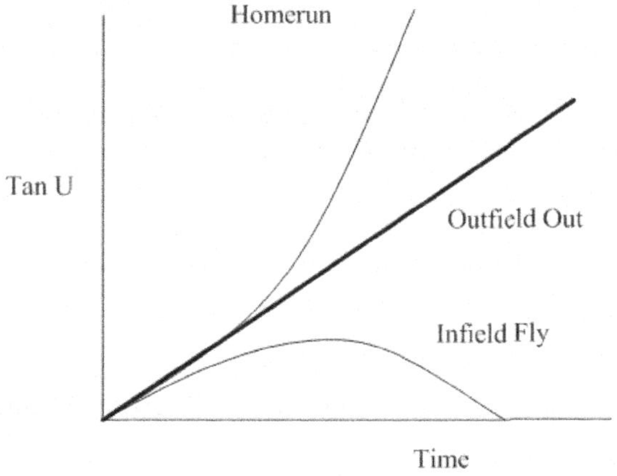

But what if the ball is hit to land 15 feet away? You might imagine that he would move quickly to the site and wait there, but he does not. Almost without fail, skilled outfielders will trot to the ball, arriving at the ball just in time to catch it.

Returning to the case where the ball is hit to land right on the outfielder: Chapman found a pattern. The tangent of the angle between the ground and the player's line of sight (looking at the ball) increases at a constant rate. If the ball will land in front of the player the tangent will increase initially, reach a peak and then decrease to 0. If the ball is going to fly over the player's head he will see the tangent increasing at an ever increasing rate going to infinity.

And Chapman made a more interesting observation concerning balls for which the player has to move to make the catch. If the player moves at a constant speed in order to arrive at the ball just in time to catch it, the exact same pattern occurs; the tangent increases at a constant rate. A player moving to catch the ball apparently adjusts his movements to obtain the pattern of the tangent increasing at a constant rate, or put less scientifically, the player moves to the ball in a manner which maintains the same pattern he sees when he doesn't have to move to the ball.

Somehow, players learn to recognize this pattern, even if they don't know what a tangent is. A baseball player moving to catch a fly ball is applying pattern recognition skills.

7.12 POWER OF INTUITION

Intuition is a powerful skill that is not well regarded in this age of reason. When we make decisions based on gut feel or intuition, we will typically conceal this fact and/or look for supporting evidence before implementing the decision. Yet, there is reason to accept a role for intuition in our decision-making process.

Pattern recognition as discussed on the previous page is a skill that is difficult to explain. The method by which we recognize faces, sentences, engine malfunctions, etc. is not well understood. The skill is largely unconscious or subconscious. Yet, we do not consider pattern recognition as odd or unusual or embarrassing.

We saw early in the course that decision making itself is not a conscious process.

Intuition occurs when we
recognize a pattern
and make a decision based on that recognition
without surfacing either process into consciousness.

Unconscious Pattern Recognition

↘

Unconscious Decision Making

↘

Answer Apparently by Magic

The fact that we are not conscious of the process is not an indictment of the results.

Story: (from Klein (3))

A fire-fighting crew entered a one-story house to fight a small kitchen fire in the back of the house. The lieutenant led a hose crew into the house and began to spray water on the fire. Rather than die down as expected, the fire roared back. They tried again with the same results. Something was not right. The lieutenant pulled the crew back out of the house to regroup. Soon after they left the building the floor they'd been standing on collapsed to reveal a raging basement fire. Had they stayed to fight they would have died.

The lieutenant in this story had no idea there was a fire in the basement of the house. He could not explain why he pulled his crew out of the house – they were not in any apparent danger. He believed he had ESP. Further introspection led by Klein Associates personnel revealed clues that the lieutenant was not conscious of at the time. The circumstances of the fire did not meet his expectations. The fire did not respond as expected to the water spray. It was too hot for a small kitchen fire, and it was too quiet for its size. The situation violated the patterns the lieutenant expected. His intuition led him to retreat from a situation that did not fit previously observed patterns.

Expert Intuition – When Can We Trust It?

Ever since the beginning of NDM research there has existed controversy between the NDM school and the Heuristics and Biases school. Recently, Klein and Khaneman have cooperated in an effort to understand where intuition works and where analytical decision making is more appropriate. Khaneman summarized the results in *Thinking Fast and Slow* (41).

In their work together Klein and Khaneman identified one particular reason they had developed different views of intuition. Klein worked with firefighters and nurses and similar professionals whose intuitions are frequently correct. Kahneman typically studies stock pickers and political scientists whose intuitions are frequently wrong. In all these fields the 'experts' expressed high levels of confidence in their predictions. The confidence level of the expert is clearly not an effective measure.

Klein and Khaneman concluded that expert intuition is likely to be trustworthy if:

1. The expert works in an environment that is sufficiently regular to be predictable.
2. The expert has had an opportunity to learn these regularities through prolonged practice.

7.13 THE POWER OF MENTAL SIMULATION

In the story of the overpass rescue (section 7.4), the lieutenant determines whether a possible solution will work by imagining the implementation of it. As he imagines the implementation of each of his first three ideas he envisions failure. He rejects each of them. When he imagines implementing the fourth solution he envisions success; so he implements that idea.

The ability to mentally simulate each possible solution permitted him to evaluate each idea without comparing it to other alternatives.

Mental simulation is used both to explain the past and to predict the future.

The ability to mentally simulate alternatives requires significant expertise. And even then, some systems appear to be too complex for effective simulation. Take economic predictions for instance. Even acknowledged experts are incapable of predicting very far into the future. Many problems are beyond our ability to simulate because we don't have an adequate mental model of the subject. Thomas Friedman, in his remarkable book *The Lexus and the Olive Tree*, discusses globalization in terms of six dimensions:

- Politics
- Culture
- National Security / Balance of Power
- Finance and Trade
- Technology
- Environmentalism

His mental model based on these dimensions allows him to make sense of the past and make reasonable predictions of the future in ways the rest of us cannot.

Working memory apparently limits simulations to about 3 adjustable factors and about 6 time steps or permutations. Experts overcome some of these limits by combining factors and steps (seeing systems as entities). The ability to see patterns is closely tied with the ability to mentally simulate.

7.14 COGNITIVE TASK ANALYSIS – CTA

Most NDM research must be conducted in field settings (an exception to this is the use of training simulators capable of generating life-like scenarios).

Studying decision making in the field can be done partially by observing decision makers in action. But this approach is time consuming, sometimes dangerous, and can never give the full picture of what is happening. Part of what the researcher wants to know is what the actor is thinking as she makes decisions. That insight can only be gained by asking the decision maker.

The most common technique used in NDM research is Cognitive Task Analysis (CTA). Many CTA techniques have been developed including the Critical Decision Method and the Decision Requirements Exercise.

Critical Decision Method (CDM)

The Critical Decision Method CTA is performed via an interview. The interview is conducted by an expert in CTA (CTA Practitioner) and the interviewee is typically a subject matter expert in the field of interest. The technique is based on the RPD method (section 7.6) and designed to identify the subtle clues that experts rely on and novices miss. The interview is conducted in 4 stages:

- Event Identification
- Timeline Development
- Detailed Assessment
- Deepening (What If Questioning)

Stage 1: Event identification

The CTA Practitioner asks the subject to identify events of interest. For example, if the study is designed to identify causes of error in engineering design, the practitioner will ask the subject to identify some occasions in which she made an error and/or someone on a project she was involved in made a significant error.

It is critical that the subject was actually involved. This cannot be a discussion of a famous error that the subject has read about, for instance. A CTA attempts to determine what went on in the mind of the person who made the error, or failed to fix it, or first noticed it, etc.

Stage 2: Establish the timeline

In the second stage of the interview the CTA Practitioner walks the subject through the event from start to finish in order to establish a timeline. The timeline identifies key decision points to be studied in detail in Stage 3.

Stage 3: Assessment of Decision Points

Stage 3 focuses on the key decisions identified in Stage 2. Skilled CTA Practitioners extract evidence in stage 3 to explain what went on in the subjects mind over the course of the event and especially at the key decision points. Of particular interest are the cues and patterns that informed the subject's situation assessment.

Stage 4: Deepening

In stage 4 the interviewer asks "what if" questions. How could events have turned out differently? "What if" the actor had been more (or less) experienced or had noticed different things, etc.

7.15 TEAM CTA

Many of the decisions we are interested in are team decisions and/or multi-team decisions. The Critical Decision Method focuses on one actor at a time. How can we use CTA to gather information about team decisions?

Klein recommends two approaches, the Team CDM and the Decision Requirements Exercise (DRX)

Team CDM:

Conduct CDM's with team members one at a time and then compare the results at each stage.

- Do the timelines match? Do team members agree on what happened when?
- Do they agree on what decisions were made and why?
- Do they agree on the important decisions?
- Do the team members agree on what cues were important? Did they all have the same information at the same time?
- Do they agree on what errors were made and how to improve next time?

Decision Requirement Exercise

The Decision Requirement Exercise (DRX) is performed with all team members present.

The participants first list the decisions made during an event and select the most critical decisions to expand.

Each critical decision is probed via these three questions:

- What made that decision hard?
- What cues or information were used to make that decision?
- What would you do differently next time?

The results are captured on a DR Table as shown below.

What Decisions Were Made?	What made this decision hard?	What information was used?	What would you do different?

The DRX is used to provide an overall understanding of team objectives, develop understanding of individual roles and responsibilities and to highlight barriers to information flow.

7.16 CTA IN LESSONS-LEARNED STUDIES

An obvious and interesting application of CTA would be as a technique for conducting lessons-learned studies on projects. CTA should help overcome a couple of limits to traditional, brain-storming type of lessons-learned studies:

In-depth vs. broad overview:

Broad-brush lessons-learned studies can tend to "learn" invalid lessons. In complex project settings it is often possible to learn different lessons from the same event depending on your point of view, access to data, etc. A more in-depth and more challenging approach such as CTA should tend to mitigate this effect.

Mitigating defensiveness:

We have seen that defensiveness (Model I behavior) limits learning. This should certainly impact lessons-learned studies when they are done within project teams. Individual CTA's, especially if conducted by practitioners with theory of action knowledge should mitigate this effect.

7.17 IMPROVING PERFORMANCE IN STRESS-FILLED JOBS

With these observations of the causes of stress and the sources of power, we can discuss ways to improve the performance of personnel in stress-filled jobs.

Long-term Memory vs. Short-term Memory

Stress has little impact on long-term memory.
Stress has a large impact on Short-term Memory

In a stressful situation, we do not forget what we already know. But stress does have a large impact on short-term memory. We have a hard time collecting data in the heat of battle and maintaining significant amounts of data readily available. Short-term memory or working memory is generally considered to be limited to about 7 items. An "item" might be a single piece of data, but it could also be a grouping of related data items, such as a profile of pressure over time. Some research suggests that we have more than one short-term memory register; in particular we appear to keep visual-spatial and auditory inputs in different lists. Hence, it may be possible to make more than 7 items available in working memory if the data transmission system is appropriately designed.

Pattern Recognition

Training in pattern recognition decreases the reliance on raw data in a crisis. Training should focus on enabling personnel to effectively assess novel situations.

Mental Simulation

The ability to mentally simulate process behavior is important both for pattern recognition and for evaluation of whether a potential solution will be effective. The training program should build mental simulation skills.

HAZOP-based Training for Pattern Recognition and Mental Simulation

A novel HAZOP-based approach for Control Room Operator training has been successfully applied by the author. A HAZOP is a structure method for identifying process hazards. In the training program the operators use the HAZOP technique to identify potential hazards. Selected hazard scenarios are then mentally simulated:

- How will this event begin (cause)?
- How will the operators recognize what's happening?
- How will the scenario progress in time and across the entire plant?
- What cues will the operators have that this is happening (especially early cues)?
- What other problems may look similar – how can operators distinguish, when will they be convinced?
- What actions are appropriate and when?
- What results will those actions have? How will the operators know they've been successful?

This training approach develops pattern recognition skills and mental simulation skills along with imparting process knowledge.

Rules and Procedures

Stress occurs with ambiguity; situation ambiguity and solution ambiguity. Some ambiguity can be and should be eliminated via rules and procedures. Well thought out procedures should be developed for non-typical operations to minimize the potential for error. For example, well developed rules and procedures that make process shutdown mandatory under certain conditions can significantly decrease stress.

Data Gathering Aids - MMI

Adequate data and a model for evaluating the data are required to establish situation awareness. Careful thought should be given to the development of man-machine interfaces (MMIs) that provide meaningful data and display the relationships between data items.

- Use tables instead of individual data items.
- Use graphics (plots) instead of tables.
- Organize data in ways that it will be used during an emergency.

- Show the operators the data they needs in a usable format while trying to shield them from unnecessary data.
- Provide calculated data in addition to measured data.
- Manage alarm generation to avoid meaningless alarms.
- Highlight data that is out of range.
- Show relationships such as pressure profiles across the process.
- Consider short-term memory limitations

Providing Calculated Data

Most control systems only show measured data; process data directly measured by end devices and transmitted to the control room. The data available for operator decision making can be greatly increased by calculating other parameters from the measured data. An example of this is virtual flowmeters. It is often possible to estimate the flowrate through a line even if no flow meter is provided. For example, pressure drop through the pipe or through a control valve or head gain through a centrifugal pump can all be used to estimate the flowrate. Other examples of providing calculated information include calculating the heat transfer coefficient of a heat exchanger and plotting a pump operating point on a pump curve.

Other Issues:

Team Building

Effective teamwork can decrease the level of stress felt by employees.

Communications Enhancement

Under stress we typically communicate less effectively with others. Communication aids can minimize this impact. In crisis management situations try to arrange it so that people who need to communicate with each other are close together and have ready access to the same data.

Consider providing common data for all to see via large display screens or computer projectors.

Data Presentation / Sparklines

Edward Tufte has developed and catalogued ways of presenting data more effectively. A particularly ingenious invention of his is the sparkline. A sparkline if a 'data word'. Here's an example sparkline of a company's revenue over several months .

7.18 RULES AND PROCEDURES

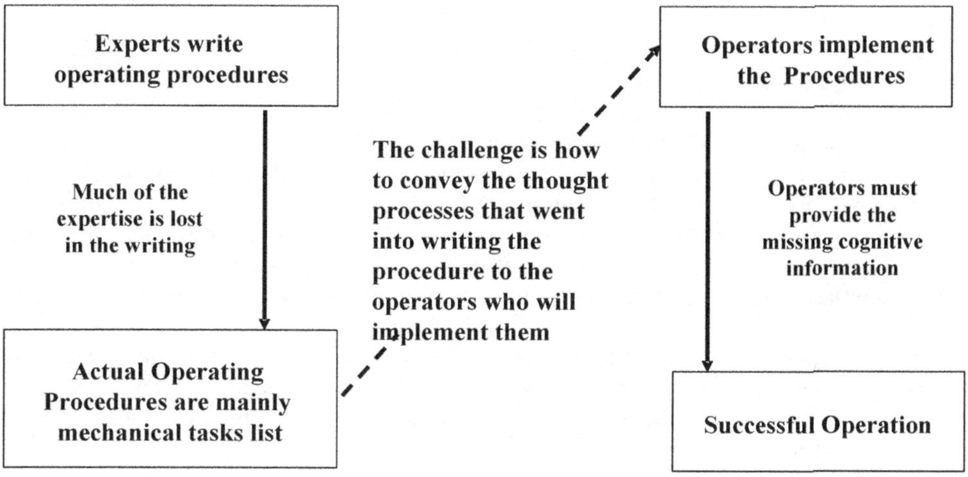

It is common practice to develop comprehensive rules and/or operating procedures to assist personnel engaged in stressful jobs. Take, for example, the operating procedures provided to chemical plant and oil platform operators. If competently written, these procedures provide detailed and effective guidelines for process startup, process shutdown and other routine operations.

The operating procedures are usually written by engineers and/or senior operators who have expertise in the process. The final result is often a list of tasks to be performed in the specified order. A great deal of expertise goes into identifying the tasks to be performed and deciding the order in which they are to be performed. But the procedures do not contain much of the expertise that went into writing them. The expertise is trapped in the experts head.

Operators implement the procedures. During normal operations, the procedures can presumably be implemented by rote and work effectively. But what if something goes wrong? Then the operator will have to respond in some way that is not captured in the procedures. A competent operator should be able to respond effectively to most problems if she understands the process and can identify what is going wrong. At this point the operator is supplying the expertise.

Even when things go according to plan the operator is providing expertise to the effort. It requires expertise and skills for the operator to implement the procedures; you could not get someone off the street to do it correctly without significant training. She will apply expertise at each stage to understand what the procedures are asking her to do. If her knowledge is different than that of the writer, she may assume reasons for some steps that are different than the reasons the writer had. This may cause her to alter or ignore certain steps that seem incorrect or spurious. Procedures written by a subject matter expert may easily be misinterpreted by an operator who is also a subject matter expert, but probably not as expert as the writer.

Two Challenges

- ***Ensuring that the author includes enough information to effectively guide a well-trained operator.***

- ***Training the operators to accurately contribute the expertise that the writer left out.***

Writing Procedures in Time

Anyone who has written operating procedures after a process has been designed (too late to make design changes) knows the wisdom of writing procedures early. The natural tendency of design engineers it to design for steady state operation in as-built condition. The writing of operating procedures for starting up, shutting down and responding to dynamic upset conditions surfaces design flaws and omissions. Operating procedures written late in the project frequently have to compensate for inadequate design.

Writing operating procedures, at least in outline form, very early in the project can also serve as an effective interface management tool. Process plants can be designed in pieces, but must be operated as a unit. Operating procedures tend to surface disconnects between design teams.

Operating procedures should be written early enough so that we can <u>design systems to be operated rather than try to figure out how to operate processes already built.</u>

7.19 CHAPTER SUMMARY

Naturalistic Decision Making is the way people use their experience to make decisions in the real world. Most decision theory research is done in university laboratories with college students as subjects. Research findings in this setting often don't translate well to the real world. Most decisions of interest are made by people with some subject matter expertise, who care about the results of the decision and who may be under time pressure.

The decision-making process employed by experts in stressful situations is fundamentally different that the typical analytical decision model. The process applied is described by the Recognition Primed Decision model (RPD).

Several skills permit application of RPD. These include pattern recognition, intuition and mental simulation skills. Expert intuition is likely to be trustworthy if:

1. The expert works in an environment that is sufficiently regular to be predictable.
2. The expert has had an opportunity to learn these regularities through prolonged practice.

NDM research is conducted mainly via Cognitive Task Analysis.

7.20 ANSWERING THE FOUR QUESTIONS

What observations from this chapter can we make about our four key engineering questions?

Why do so many changes occur late in projects?

Poor mental simulation of the design:

We tend to design for steady state operation in as-built condition. Mental simulation of other operating conditions (startup, shutdown, upsets, equipment failures, etc.) would identify design flaws earlier in the project.

Focus on plausible objectives:

Experienced engineers can predict which objectives will be easy to achieve and which will be difficult. There is a tendency to focus on achievable objectives and to dismiss those which are difficult to achieve. Pursuing difficult objectives late in the project causes design changes.

Why do we have so many problems at interfaces?

Decisions made easily and naturalistically by one team may be difficult to sell to another team with less expertise or different expertise:

We make many engineering decisions naturalistically; we quickly recognize a situation and apply solutions we know will work without identifying and evaluating multiple alternatives. Selling such solutions to other teams or to management can be difficult. In areas where others lack expertise they may expect analytical decision making.

Why do we repeat past mistakes?

Harder to introspect for errors in naturalistic decisions:

It may be more difficult to recognize errors in decisions made naturalistically. Consider a case where you actively considered several different alternatives before selecting one that didn't work. In this case you have ready fodder for learning. It is easy to ask "what if I had chosen option __ instead?". In the case of a decision made naturalistically you may not have considered any other options which might have worked and hence will not have ready access to a feasible solution to compare to the one you chose.

Why do we so often finish projects behind schedule?

No significant learnings from this chapter.

Chapter 8
Teams and Interfaces

Sources:

Much of the insight for this chapter comes from *Social Cognition* (30) by Susan Fisk and Shelley Taylor.

Another important source is *Group Process, Group Decision, Group Action* (31) by Robert Baron and Norbert Kerr.

And finally insight into factors important in making teams work comes from Gary Klein (3).

Preface:

Teams occasionally achieve great results, but they frequently don't work very well. And in organizations composed of multiple teams we frequently have problems at the interfaces between teams. Animosity between teams, or "silo" behavior, is prevalent and is a significant cause of business underperformance.

Creating teams creates the opportunity for teamwork, but creating teams also creates turf.

CHAPTER 8 CONTENTS

8.1 INTRODUCTION TO TEAMS
8.2 TEAM BASICS
8.3 TEAM FORMATION, TEAM CULTURE, THEORIES OF ACTION
8.4 TEAM COMPETENCE
8.5 BUILDING SHARED MENTAL MODEL TEAMS
8.6 NEGOTIATION – THE GREEK WAY
8.7 EVALUATING INTERFACES
8.8 INTERFACES BETWEEN ENGINEERING TEAMS
8.9 IDENTIFYING OBJECTIVES
8.10 DATA SHARING
8.11 EFFECTIVE COMMUNICATION DURING SYNTHESIS & ANALYSIS
8.12 TESTING
8.13 THE BRAIN AS A METAPHOR FOR PROJECT ORGANIZATIONS
8.14 CHAPTER 8 SUMMARY
8.15 ANSWERING THE FOUR QUESTIONS

8.1 INTRODUCTION TO TEAMS

Most meaningful engineering tasks or projects involve the collaborative effort of multiple people organized into a team and/or into a project organization consisting of multiple teams. Where work is done by teams of people and groups of teams we have interfaces. Effective management of these interfaces is critical to the effective execution of projects. As projects become larger, more complex and more geographically dispersed the interfaces become more important and more difficult to manage.

Reasons for Teams

An individual actor who has adequate skills, knowledge, time, authority and motivation will usually perform more efficiently and effectively than any team. Work is performed by groups of people (teams) for one or more of the following reasons:

Work Quantity:	the quantity of work to be done exceeds the capacity of a single actor.
Work Diversity:	the diversity of the work is such that no single available actor has all the necessary knowledge and skills to adequately perform all the required tasks.
Stakeholder Diversity:	stakeholders with different objectives and allegiances must be, or desire to be, represented on the team.
Inclusion Culture:	teams may be constructed simply because of a business culture favoring inclusion even when there is no valid business reason.
Social Needs:	we have a social need to belong to groups.

Division of labor driven by work quantity considerations suggests formation of a functional team such as in a team of process engineers. The members of a functional team share the same general skill sets, Each member of the team will be doing similar work.

Division of labor driven by work diversity considerations leads to formation of a multi-disciplined team consisting of at least one member with each required knowledge/skill set.

Some team membership is driven by stakeholder's perceived need to safeguard their interests. For instance, it is common to include Operations representatives on design teams to insure that Operation's objectives and concerns are voiced.

8.2 TEAM BASICS

Much of the material in this section is taken from *Group Process, Group Decision, Group Action* (31) by Robert Baron and Norbert Kerr.

Groups play a strong role in our lives. We want to belong to groups. From infancy we learn to depend on others. Our self-esteem depends on how others perceive us. But the need to belong is balanced by the need to be unique and special (30) and to be in control (1). When groups make us feel anonymous we may resist conforming to group norms.

Team Productivity

The literature contains stories of "skunk teams" whose work output and creativity far exceed what the individual team members would otherwise have been capable of. Even if these are not simply urban myths, it is unrealistic to expect such performance from teams typically. Teams will typically be less efficient than individuals. By understanding the factors at work, we can make educated attempts to improve team productivity.

Team potential is a function of the members skills and resources. The actual work output of the team is its potential minus process losses (31):

Actual Productivity = Group Potential – Process Loss

Causes of Process Loss

Coordination Losses:

- A group may fail to solve a problem even if one of the group members can solve the problem because she must communicate her solution to the team and convince the team that her answer is correct.
- Individual team members may work on spurious activities not directly related to team mission, and/or may duplicate the work of other if the team objectives and roles and responsibilities are not clear.
- Teams meet, communicate, coordinate, argue, etc. Constraints or distractions arise from the process of interaction:
  - Production blocking – only one member can speak at a time, others have to wait their turn. Some thoughts are never voiced as a result.

– Evaluation apprehension – depending on the level of trust within the team, some members may not fully share their thoughts for fear of embarrassment.

Motivational Losses:

- We like to be in control (Model I behavior). Groups diminish our personal control. In important situations where we disagree with the group consensus, we will be less motivated to do our share. In particular, if the team leader is dictatorial we should expect large motivational losses.
- Groups create a shared sense of responsibility. The responsibility that each individual senses is lowered by group membership. Individuals feeling a low sense of responsibility are more likely to loaf.
- If it is known that other group member(s) will take up the slack then loafing by some is more likely.
- If it is not clear what impact individual effort will have on the group product then individuals are more likely to loaf.

Both coordination losses and motivation losses tend to increase with group size often in a non-linear fashion. To get a sense of the impact of group size on coordination losses consider the number of possible interfaces between group members as a function of group size as shown on the plot below. The dashed line is the 45 degree line (X = Y). For teams with up to five members the number of interfaces is about equal to the number of team members; for a ten member team the number of possible interfaces with others is 47, five times the number of team members. So we should expect coordination losses to grow exponentially with group size. At some point coordination losses will exceed the potential benefit of adding team members – adding members will actually decrease the work output.

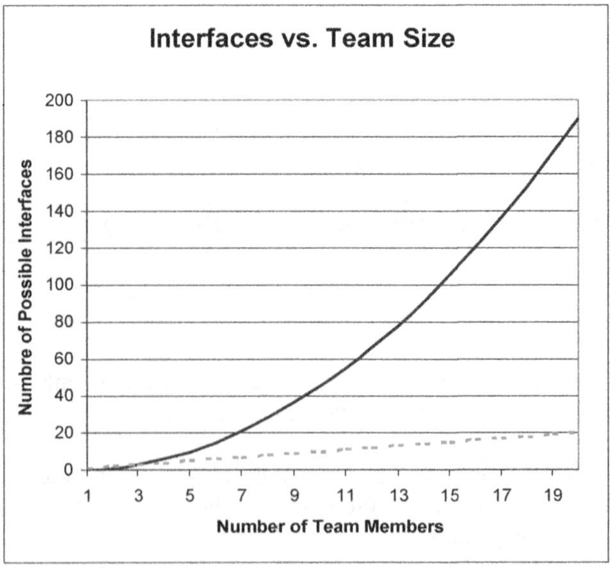

Klein (3) reports an example in a study of an emergency response team with over 80 members that was performing very poorly in drills. When the team was cut to fewer than 40 members and efforts made to improve coordination between members the emergency drills went smoothly. Coordination losses made the larger team less effective than the smaller team.

Social Influence and Conformity

Group membership tends to create conformity of views. There is social pressure to accept the party line. There are distinct benefits to this behavior: the group consensus opinion is usually correct and group consensus is important for coordinated action. Even when the group is wrong at least consensus leads to action and action is usually preferable to inaction. But this pressure to conform is demoralizing to those who disagree with the team's opinion.

Intergroup Conflict and Aggression

Teams create turf. Turf must be defended. Teams create an Us vs. Them mentality. Simply creating a group may be sufficient to trigger intergroup hostility and discrimination.

There are a number of reasons why we will favor our own group including:

- Favoring one's own group reflects a need for self-esteem. (We feel better about ourselves because we belong to the better group.)
- We favor our in-group because it is expected of us.

This favoring of the in-group results necessarily in our forming a lower opinion of out-groups. We view in-group members as unique individuals, but out-group members as being "**all alike, different from us and bad besides**" (30).

Aggregation of Team Knowledge and Polarization

It seems logical to expect that a team's opinion on issues would be the average of the individual member's opinions, but that is not always the case. In some situations groupthink polarizes opinions such that the group makes decisions that are more radical than any individual team member would have made. To a large extent this behavior is determined by the way individual views are expressed and aggregated into a team view.

In situations where the teams contain significant diversity, or where at least one person is willing to take a stand in opposition to the majority view, teams tend to moderate the opinions of individual members.

But in very homogenous teams, and in teams where no individual member is willing to challenge the majority view, discussion within the group may become more and more polarized. Groupthink can lead to a polarization of views such that team decisions are more radical than any individual member would have selected.

Learning

The Norse colony in Greenland was founded in the 900's by the Viking Lief Erickson. The colony prospered initially, but died out after about 450 years with all inhabitants probably having starved to death. An amazing aspect of them starving is that they did not eat fish! We know this because extensive archeological investigations have discovered no fish bones in the trash heaps of the Norse. While it is not unusual for different cultures to have different tastes in food, this oddity was newly learned. The Norse ate fish everywhere else they lived. Somehow they must have "learned" that fish in Greenland were not edible very soon after arriving in Greenland.

We will likely never know how the Norse came to view Greenland fish as inedible. From a concept of team learning what is interesting is that for 400 years every Norse believed it even though they must have seen the local Inuit enjoying meals of fish while they went hungry. And they must also have had to repeatedly explain the lack of fish in their diet to Europeans visiting on trading ships.

The Norse competed on Greenland with the Inuit who survived to modern times. The Inuit's knew many things about surviving in Greenland including successful methods of fishing for seals and whales and the use of seal hides rather than scarce lumber to make boats; techniques that the Norse never learned.

The Norse experience illustrates two things about the impacts of teams on learning:

- It is difficult to learn something that is at odds with your team's preconceived opinions.
- It is difficult to learn something from other teams viewed as inferior.

As a group the Norse were unable to learn from the Inuit because they considered them to be inferior. As individuals, the Norse were unable to learn things that the rest of their group did not believe.

While teams can facilitate learning they can clearly also impede learning. We saw in chapter 6 that Model I behavior impedes learning. Teams accentuate Model I behavior and hence we should expect teams to be more resistant to double-loop learning than individuals are.

8.3 TEAM FORMATION, TEAM CULTURE, THEORIES OF ACTION

The key insights for this section are from Schein (36) and Argyris (1).

What is a team? A simple, but effective answer is that a team is a crowd with culture.

Crowd	+	Culture	=	Team

This answer, of course, begs the question "What is culture?" Schein (36) defines group culture as:

> *A pattern of shared basic assumptions that was learned by a group as it solved its problems of external adaptation and internal integration, that has worked well enough to be considered valid and, therefore, to be taught to new members as the correct way to perceive, think and feel in relation to those problems.*

By "shared basic assumption" Schein is referring to deeply held values that are taken for granted and are nonnegotiable. Schein discusses culture as occurring at three levels:

1. Actions

2. Espoused Values
3. Underlying Assumptions

If the term "espoused values" hasn't grabbed your attention you didn't pay attention during the theory of action discussion in Chapter 5. This model for team behavior is analogous to the model for individual behavior.

Model II – Espoused Theory	Model I – Theory-in-Use
non-threatening situations not much at stake	threatening or important situations great deal at stake
• **Try to be fair** • **Seek and value valid data** • **Involve other stakeholders in decision making**	• **Try to win, to be in control** • **Seek data that supports your objectives** • **Act unilaterally in ways that inhibit inquiry** • **Try to save face, avoid negative feeling**

From this model we should expect teams to:

- Espouse respectable, upright values and act according to espoused values when nothing important is at stake, but
- Also hold more important values that are largely tacit and unacknowledged. These values will guide behavior when something important to the team is at stake.

A teams espoused values are easy to identify. They will be stated in team documents such as the vision and mission statements. They will be preached by management when pep talks are necessary. They will be shared with interviewees and new hires. But we should not be surprised to find that teams violate their espoused values in times of crisis or perceived crisis.

Model II Behavior within the Team - Model I Behavior towards Outsiders

Behavior within a closely knit team should be largely per Model II theory (espoused values). Indeed, a strong culture can have a remarkable positive impact on individual behavior within the team. Selfless behavior (taking one for the team) is highly valued. Members will even sacrifice their lives for the team in some situations involving religious and military groups.

The most beneficial impact of culture may be this tendency for Model II behavior within a team.

But we cannot expect such behavior to cross team boundaries. In most situations we should expect the opposite; that team formation will increase the tendency toward Model I behavior towards outsiders. In *Culture Matters,* Huntington and Harrison, note that societies in which families ties are the closest tend to have the highest levels of government corruption (strong in-group bonds yields weak out-group bonds).

How can we know what values a team holds most dear and, how do team members learn them? Team members learn these values by experience. If an organization encourages its employees to strive for work-life balance, but promotes only the workaholics to management positions; employees learn that value quickly enough.

Team Formation and Stagnation

A group learns its culture by applying ideas that are successful. Often these ideas are the ideas of the group founder. The ideas may be open to discussion initially, but once they are applied successfully they become part of the groups fundamental assumptions and become non-negotiable. Effectively, anyone who doesn't accept the shared basic assumptions cannot be part of the group.

A team with a very strong culture may come to think that its way is the only way, that it has little to learn from others and hence, to lose the ability to learn and adapt.

8.4 TEAM COMPETENCE

One of the areas of interest to NDM researchers is the impact of teamwork on decision making. The model of team competence below is a simplified version from Klein (3).

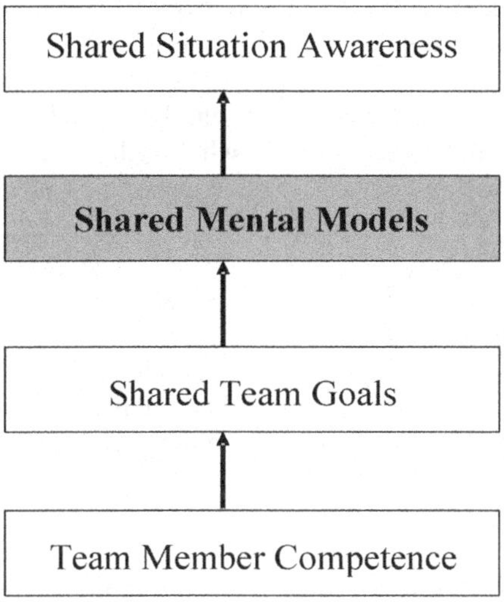

Does everyone agree on what the current issues are? Do the team members trust each other? Do the team members have a realistic view of how the team is doing?

Does every team member understand the role of the other members so that he/she can interact effectively with others?

Do all team members understand and support the overall team goals?

Are all individual team members competent to perform their roles and do they know what their roles are?

Individual Competencies

We usually judge teams by the talent of the individual members. Teams can perform competently only if the individual team members are competent at their jobs and know what their jobs are.

Assessment: How good are individual team members at their jobs?

Team Identity (Shared Goals)

Individual competencies are important, but a team composed of superior performers can still perform poorly if the members don't understand the objectives of the team. The real benefit of teamwork begins when everyone begins pulling in the same direction towards a common goal.

Building a team identity takes time. In new teams, or poorly organized teams, each individual may simply be doing his/her job without much regard to the whole scope.

Having a strongly shared goal is probably the most effective ingredient in building a team.

Assessment: Does everyone know the team goals?
Does everyone agree with the team goals?

Team Cognition (Shared Mental Models)

Understanding the team goal is a big step. The next step is for team members to share a common mental model of the work to be done and the way it should be done. Each team member should understand what the roles of other team members are and how the decisions and actions of each team member affect others on the team.

This level of understanding allows more effective interaction between team members. A team member contemplating a change will understand the impact the change will have on other team members. Team members will alert each other to changes in one area that impact others. Data important to others will be shared before it is requested.

Shared mental models allow teams to adapt to changes more quickly and effectively because they will have a sense of how their teammates will react, what each person's role should be, who needs what information, etc.

The team is beginning to think with one mind.

Assessment: Does everyone have the same picture?
Does everyone understand how their work impacts others?
Does everyone know how other peoples work affects them?
Do people help each other out vs. focusing only on their own tasks?

Team Metacognition (Shared Awareness)

The degree to which team members have the same interpretation of ongoing events is important for team performance. In dynamic situations it is easy for team members to develop different views of what is happening without realizing it. Team members may continue working towards objectives that have been abandoned by others or continue using data that has been superseded.

Sharing all available data is not the answer – that only creates data overload. The key is control of attention. Teams must have access to important data. But simply making data readily available does not guarantee that the people who need it will read it and understand it. We all typically focus on the work directly under our control and much less so on the work of others, especially others on other teams. Even if reports are sent to all the right people we cannot assume that those people will read them and understand them in the way that the author meant.

Effective teams make sure that relevant, important information is effectively communicated and attended to?

Another important factor in team meta-cognition is whether the team can evaluate its own performance and know when it is in trouble and must hence either ask for help, change course or perhaps simply provide more support to a team member who is struggling.

Assessment: Does everyone agree on what the current issues are?
Does everyone share the same opinion on how effectively the team is working?
Do the team members trust each other?
Does everyone pitch in to help a colleague who is struggling?

8.5 BUILDING SHARED MENTAL MODEL TEAMS

The concept of share-mental-model teamwork is the key observation of this chapter for improving interfaces between teams. It is at this level that teamwork pays large dividends. At this level people can coordinate work without large efforts. Communication is simplified and more effective.

How do we build shared mental model teams? How do we build a shared-mental-model level of competence across an entire project organization?

Crosstraining - Shared Knowledge

It is common for us to spend most of our professional development time and effort on improving our expertise in our area of specialization. But, since our errors are largely at the interfaces, a significant fraction of our professional development effort should be spent on learning the fundamentals about what others do with the goal of facilitating meaningful communication between groups.

Experience-based Approach to Crosstraining

One approach to cross-training is to identify the types of interface problems that have occurred and/or might occur and to identify from these events the types of knowledge required.

Teach Simple Heuristics

Another approach is suggested by the work of Bill Koen. In *Discussion of the Method* (33) Koen describes the engineering method as (paraphrasing):

> **Engineering is the use of heuristics to find the best solution to an ill-defined problem subject to resource constraints.**

Koen observes that all of engineering is the application of heuristics. We have no simple fundamental truths; all design techniques are approximations. All techniques we use are heuristics. We cannot teach others how to do our jobs using the heuristics we use. That would take far too long. But we can teach people in other teams the simplest set of heuristics that we could use to do our jobs.

For instance, I have taught an oil separation system (topsides) process design course featuring only about 12 simple heuristics. These can easily be taught. The class then uses those heuristics to "design" a topsides process as part of the course. Any engineer taking the course has actually designed a topsides separation facility via simple heuristics and should then be much better equipped to interface with topsides engineers.

8.6 NEGOTIATION – THE GREEK WAY

When there are multiple stakeholders and differences of opinion, negotiation may be necessary to achieve agreement. Much of what we've covered in the book may be useful for constructing an effective argument. Insights include:

1. Consider every stakeholder's objectives. Focus on shared values and objectives.
2. What data and information do the stakeholders have access to?

3. Be conscious of Model I behavior. Effective agreement is much more likely if all stakeholders are behaving per Model II theory of action.

These and other insights from the decision theory may help you structure effective negotiations and improve the odds of reaching an effective decision. It is also valuable to study and apply the Greek art of rhetoric.

Rhetoric

The ancient Greeks were serious about the art of arguing, or what they called rhetoric. The Greek's perspective on negotiation is succinctly and humorously conveyed in a short book titled *Thank You for Arguing* (40) by Jay Heinricks.

Rhetoric identified three tools of persuasion:

- Argument by character (ethos).
- Argument by logic (logos).
- Argument by emotion (pathos).

If the facts are on your side, then it makes sense to argue via logic (logos). Stick to the facts. Challenge others who misuse the facts. When you challenge other people's arguments look for the following fallacies:

- False comparison – two things are similar in some way so they must be similar in other ways (representativeness heuristic).
- Bad example – the example is false, irrelevant or wrongly interpreted.
- Lack of evidence used as evidence of lack.
- False choice – not all of the options are given.
- Red herring – an irrelevant argument used to distract.
- Wrong ending –"proof" fails to lead to conclusion.

If the facts do not fully support your position, then argument by character (ethos) may be more successful. And even when arguing by logic, argument by character should be employed first. Any argument will be more persuasive if you have the audience's trust. There are three essential qualities of persuasive ethos:

- Virtue – the audience believes you share their values.
- Practical wisdom (street smarts) – you appear to know the right thing to do on every occasion.
- Selflessness – the audiences interest seem to be your sole concern.

Argument by emotion may be the most effective method if you want to convince the audience to take immediate and decisive action. When arguing by emotion, consider the tense. Use the tense appropriate to your goals:

- Past tense to place blame: If you want to place blame speak of the past.
- Values (present): To convince the audience of shared values speak in the present tense.
- Choice (future): To motivate the audience to make a decision focus on the future (how the decision will impact the future).

Carefully monitor the arguments of your opponent and challenge these fouls commonly made during argument by emotion:

- Switching tense away from the future.

- Inflexible insistence on the rules, refusing to hear the other side.
- Humiliation.
- Innuendo.
- Threats.
- Nasty Language.

8.7 EVALUATING INTERFACES

Interfaces within a Functional Team

The easiest interface is one between engineers working within the same functional team, such as two process engineers working within a process engineering team. Several factors make this interface relatively easy:

- they will be working on similar tasks,
- they will work with largely the same data set,
- they will possess the same basic skill and knowledge set,
- they will likely share the same set of design objectives,
- they should recognize easily how decisions they make impact each other, and
- they will typically be located at the same site and will usually know each other personally.

Interfaces within a Multi-disciplined Team

An effective way to staff a small project is via a multi-disciplined team. For instance, a small project team could consist of a mechanical engineer, a process engineer, a control systems engineer and a civil engineer working together to generate the design. Here the interfaces are a bit more challenging. The interfacing parties:

- will be working on different tasks,
- will be using different data sets (with some overlap),
- will not share the same knowledge base,
- and though they should share overall project objectives, they will have some different, possibly conflicting objectives.

Interfaces between Functional Teams

Large project organizations are frequently composed of multiple functional teams. The interfaces between such teams are particularly challenging. At this point we must ask the meaning of an interface between teams. Indeed, no such thing really exists. Interfaces exist between individuals only. The interface between teams is the collective impact of interfaces between individual team members.

These interfaces are similar to interfaces within a multi-disciplined team with one or two very large differences:

- the functional teams may not be located at the same site, and
- the individuals interfacing are members of different teams.

The second issue is crucially important. Team membership is an important social force which impacts the quality, type and timing of information transfer. Significant social and cognitive factors must be accounted for if we are to successfully understand and manage the interfaces between members of different teams.

8.8 INTERFACES BETWEEN ENGINEERING TEAMS

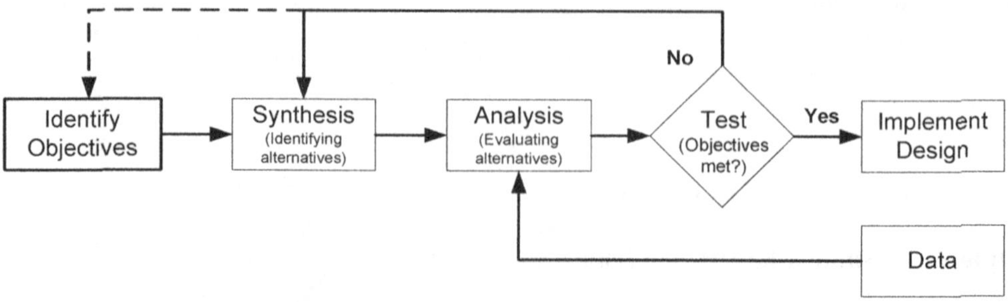

The graphic above is a simplified engineering model:

- The model begins with identification of objectives.
- Having identified objectives we attempt to identify design alternatives which will satisfy those objectives (synthesis).
- Each promising design option is evaluated against the design objectives (analysis).
- We then ask whether the design option satisfies all of our design objectives (test). If the design option satisfies all of the identified objectives we may proceed to implement the design option (detailed design, construction, etc.).
- On the other hand, if the design option under consideration does not satisfy one or more of the identified objectives we return to synthesis and try to identify a better alternative.
- The dotted line suggests that our objectives change over the course of the design. The design effort reveals some objectives we initially missed. And if we fail to identify an alternative which satisfies all of our objectives, then we will eventually be forced to relax one or more of our objectives.

Over the course of a project this model is applied repeatedly, more or less rigorously, to each of many, many design issues.

Where multiple teams design concurrently we have the situation depicted below.

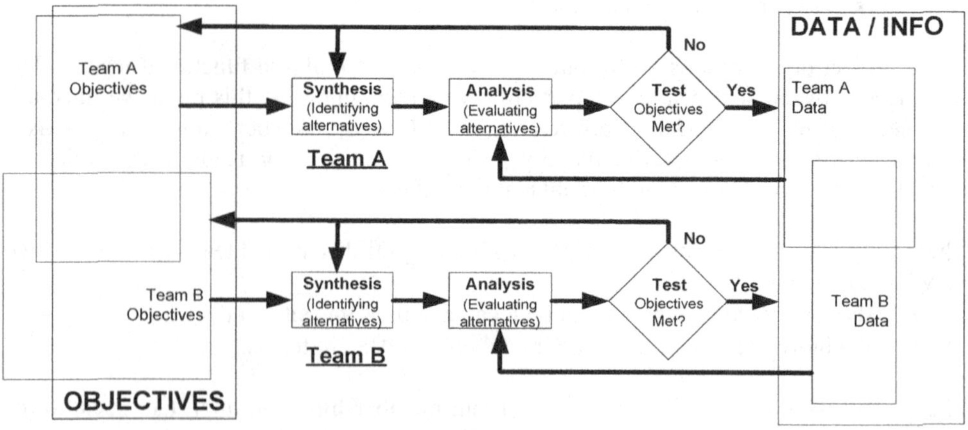

Here we see two teams concurrently designing separate parts of a facility. Several aspects of this sketch are of interest to us:

Objectives: The solid rectangle labeled "OBJECTIVES" represents all the valid project objectives (admittedly an ambiguous concept). Here we see each team designing to a subset of the objectives, and we see each team possessing some objectives which are not valid project objectives.

Data: The other large rectangle labeled "DATA/INFO" represents all the data and information available to the design teams. Note that the design efforts of each team generate information then available to the other teams. Each team is using a subset of the available data.

Testing: After analysis, each design team asks itself whether the alternative under consideration adequately meets all of the design objectives. The objectives should include all the valid objectives for the project including the objectives of other teams. But are the individual teams competent to evaluate their design against the objectives of other teams?

Communication during Synthesis and Analysis: The drawing suggests that the only communication between teams is about objectives and data. But some communication during the synthesis and analysis stages is important, especially when developing novel designs with significant interaction between systems.

Keys to Effective Interfaces

This view suggests that the keys to effective interface management are:

1) Effective identification and sharing of objectives,
2) Effective sharing of data, where much of the data of interest is generated by the design efforts of the teams,
3) Effective communication between teams during synthesis and analysis as occasions arise which demand such communication,
4) Effective testing of individual team designs against the overall project objectives.

The remainder of this chapter focuses on those issues.

8.9 IDENTIFYING OBJECTIVES

Why Identifying Objectives is Difficult

We saw in Chapter 3 that identification of objectives is difficult for a number of reasons including:

Ambiguity: There is no correct set of objectives for a given project. We do not simply identify the project objectives, we choose the objectives. Different engineers working on the same task will choose different objectives or put different weights on the objectives. As a result a design that is satisfactory to one engineer may be unsatisfactory to another.

Plausible Objectives Selected: When an experienced engineer starts a design she knows generally what it will be possible and practical to achieve. As she sets the objectives for a design she will tend to select objectives that are more easily achievable and to reject those that will be more difficult to achieve.

Constantly Changing: Our objectives are not constant; they change over the course of the project. These changes result from factors such as: late identification of objectives which ideally should have been identified earlier, failure to identify a design option which meets all of our objectives, identification of a design option which has a beneficial feature which we had not initially hoped to achieve, new people join the project and express different goals, etc.

Alternatives-focus: We are action oriented. Management values progress. Our tendency is to quickly begin identifying and evaluating alternatives and, upon identifying a viable option, to quickly begin implementing that alternative. In this way we show rapid progress – though often it is progress in the wrong direction. Focusing on objectives does not come naturally to us and even when we do explicitly identify objectives we tend to do it incompletely.

Tacit Objectives: We've done projects many times before. We assume that we know the objectives for the project; that they will be the same as the objectives for the previous ones. Many of our design objectives are tacit and are never made explicit.

Hidden Objectives: Teams and individuals may hold some objectives which are not valid project objectives and which they may therefore keep hidden. Reasons for this can include:

- Work quantity (team wants to do more work or less work than the project wishes to assign to them),
- Team promotion and growth (team wants to steer alternative selection to an area where they are most proficient and will excel and/or away from an option requiring skills that they do not possess),
- Principled objection (team may disagree with project management for valid technical or ethical reasons),
- Emotional reasons (such as preference for or against a particular vendor because of personal relationship with a salesperson).

Any of these motivations may be difficult to discuss explicitly.

Different Stakes for Different Stakeholders: Different teams will naturally have some different objectives, some of which may be mutually exclusive.

Success Criteria

In order to have effective interfaces we must insure that:

- The overall project objectives are adequately identified,
- Each team's objectives are consistent with the overall objectives,
- Each team understands the objectives of other teams,
- None of the decisions of one team prevent another team from meeting its objectives.
- The project environment encourages the surfacing of hidden objectives (may be too much to hope for), and
- Changes to objectives are communicated effectively.

8.10 DATA SHARING

It will not surprise anyone that effectively sharing data is an important aspect of interface management. Some interface management programs are based almost exclusively on sharing data. But there is a great deal more to data sharing than first meets the eye.

Databases, Drawings, Reports, etc.

It is obviously important that data be made available to those who need it. This is usually done formally via databases, drawings and reports. It is important to get this right. It is also important not to overlook the informal pathways for data flow. Much of the information that people need they get by visiting, calling or emailing someone who knows. Data sharing is improved by improving relationships between people on the project.

It is important to avoid ambiguity in data. For instance, the term "design pressure" is easily misinterpreted. Depending on the context and on the team doing the interpreting "design pressure" may be interpreted as the normal operating pressure, the maximum feasible operating pressure (PSH setting), the maximum allowable operating pressure (PSV setting) or even the maximum incidental pressure or the hydrotest pressure.

Theory of Action – Model I Behavior

Effective sharing of data requires that people <u>want</u> to share valid data. But the more important the data is the more difficult it may be to share it. Model I behavior causes us to resist sharing data that is embarrassing or that does not support our team's point of view. At best there will be a delay in transmitting such data; the data may be shared with other teams only after it has been thoroughly dissected and the best possible spin put on it.

Structural Secrecy

The structure of organizations creates structural secrecy. Individuals collect data, evaluate it and generate inferences from it. They pass their inferences on to others. The raw data is rarely shared outside of a small group of specialists. When we infer the meaning of the data we collect, we will tend to infer meanings that support our objectives and our team's objectives. The inferences of one team then become the "data" for another team. Each team will make some inferences from the "data" and pass these inferences on to other teams and up to their management. As we get further and further from the data the conclusions drawn will be based more and more on inferences by people without specialized expertise.

We tend to assume that communication is about sharing data and informing others. Sometimes it is, but usually it is to about conveying meaning.

We communicate data to inform.
We often communicate information (meaning) to control.

Control of Attention – Timely Exchange of Information

Information has time value. Simply making data readily available does not guarantee that the people who need it will read it and understand it and respond to it in a timely manner.

Communication – Why it's not the Answer

Attend almost any panel discussion on improving engineering processes and you will be encouraged to "Communicate, Communicate, Communicate." This seems like such obviously good advice that it is never challenged. But this advice tacitly assumes that what is communicated is valid and accurate data, and that it is transmitted simply to inform. But that is likely to be true only in non-threatening, relatively unimportant situations. When the communication is about relatively unimportant things it takes the form of conversation. When important issues are at stake the communication assumes the form of argument. Each party tries to convince the other party. Defensiveness peaks in important situations when a great deal is at stake. At the very times when communication is the most important it is the most inhibited. Increasing the quantity of communication without tackling the issues that make communication ineffective will not improve project performance.

Another reason that additional communication may not be effective is the difficulty in commanding attention. Communication is only effective if we get the attention of the other party and that party has the knowledge and situation awareness to understand the communication. We are fooling ourselves when we send an email with a significant message and expect that the receiving party will read it, understand it, apply the same meaning that we intended, and appropriately act on the information.

8.11 EFFECTIVE COMMUNICATION DURING SYNTHESIS & ANALYSIS

The effective sharing of objectives and data, though not necessarily easily done in practice, at least is conceptually straightforward. The objectives can be identified via structured methods and shared via reports. The data can be shared via reports and databases.

But effective communication during the design effort (synthesis and analysis steps) is a much fuzzier concept. Some items requiring coordinated design effort can be identified a priori and accounted for, but many items requiring coordinated action simply occur in the midst of design. How can we accommodate these? Consider this example:

An availability study of an oil production platform has suggested that the platform will be down for significant periods of time due to gas dehydration system outages. This is so partly because the gas export line is specified without any corrosion allowance. To avoid corroding the line, production must be shutdown whenever the water concentration in the gas exceeds the spec. It is too late in the project to consider adding corrosion allowance to the gas export line design specification.

But suppose that the gas export line laying stresses are the limiting case for the majority of the pipeline routing in deep water. In shallower water the pressure rating dominates. Hence, the flowline contractor's plan is to lay thicker wall pipe through most of the run to accommodate the laying stresses and thinner wall pipe (consistent with the pressure rating) in the shallow water section where the laying stresses are lower.

If the thicker wall pipe is used throughout the topsides engineer will get the corrosion desired allowance. The flowline contractor has no driver to do this, of course. His mission is to install the line within specification at the lowest possible cost. But the cost increase may be easily justified by the new topsides objective, and may be easily implemented if the two teams recognize the opportunity.

This is an important opportunity to interface. It is easy to imagine it being missed. It is also easy to imagine the conditions under which the conversation will take place:

- If there is a close working relationship between the two teams the issue may be raised. It may even occur by accident if there is ongoing contact about other issues.
- The more that engineers on each team understand the objectives and constraints of the other team, the more likely it is that they will realize that an opportunity exists.

In a phrase, this opportunity is most likely to be realized if the parties have a shared-mental-model level of awareness across teams.

8.12 TESTING

The question "are the objectives met?" is more difficult to answer in the situation where multiple teams are working concurrently. Teams need to judge not only whether their design meets their team's objectives, but also whether their design meets the objectives of other teams. The tests conducted by individual teams will more likely rise to this level if they are at least aware of the overall project objectives. But more is required; design teams cannot surrender control of their objectives to other teams and must be involved in verifying that other group's designs are adequate. Project management must have processes in place to facilitate the timely review of all work by all impacted stakeholders.

8.13 THE BRAIN AS A METAPHOR FOR PROJECT ORGANIZATIONS

I close this chapter with an interesting metaphor for teams. The human brain is a very interesting metaphor for studying project organizations. (The reptile brain might be a better model for some organizations I've seen.)

The human brain is divided into multiple areas. There is a hierarchy of functions, some areas in control; other areas subservient. There are specialized areas such as systems that specialize in sight. Specialized areas depend on input and support from other areas and systems to complete their tasks. Areas of higher understanding (management?) depend on input from multiple specialized systems (hearing, sight, touch, etc.)

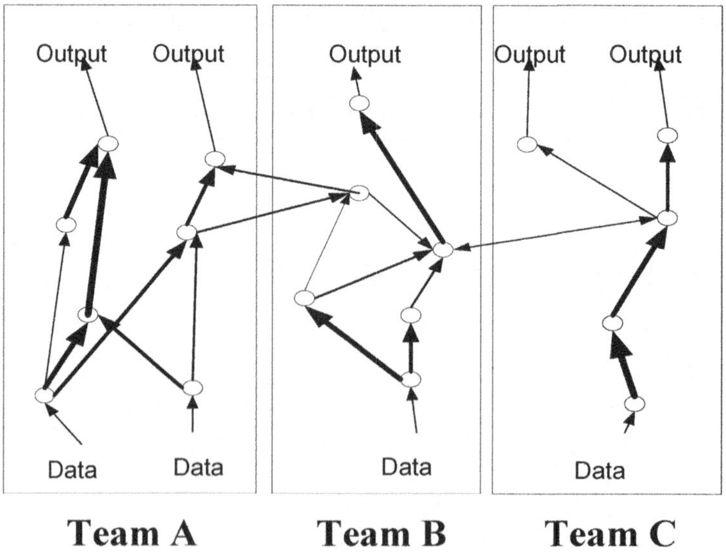

Neurons and Synapses

The circuitry of the brain consists of neurons and synapses. Neurons are connected by synapses. Neurons "fire" when stimulated. When they fire they send signals to connected neurons. The impact of one neuron on another depends on the strength and sign of the connecting synapse. Synapse connections can be positive or negative. Hence the firing of a neuron can impact connected neurons strongly or weakly and positively or negatively.

The network generates outputs in response to input data as some overall function of the connectivity of the network.

Synapse connections are strongest within specialized areas of the brain and weakest between specialized areas.

All of this mimics engineering organizations. Relationships are strongest within teams. Connections between members of different teams are more dispersed and tend to be weaker. Some connections are positive (you are likely to support what they propose) and some are negative (you are likely to resist what they propose).

The graphic above shows a few neurons with a few synapses. Consider the three areas to be teams. Connections are strong within each team, though they could be stronger and more uniform. Connections between teams are weak and rare.

General Observations

Response of an organization to data and changes in data is a function of connectivity between the players. This is especially pertinent in cases when data surfaced by one team is important to another team. How rapidly does data and understanding permeate through your organization?

We have discussed interfaces between teams. But there is no such thing. The interfaces are between people, some of whom are on different teams.

Most organizations pay most of their attention to neurons (individuals with skills). Synapses are as important or more important. How do we improve synapses? There are several obvious ideas, some of which we've already discussed:

- Cross-discipline training to foster understanding.
- Data and Knowledge Management Systems.
- Management initiative to explicitly identify the people who need to know each other and work together.
- Teambuilding events.

8.14 CHAPTER 8 SUMMARY

Team Productivity

Team Productivity = Team Potential – Process Losses

Process losses include coordination losses and motivation losses.

Coordination losses increase non-linearly with team size. Hence, we should expect small teams to be more efficient. We should expect that at some point coordination losses will dominate and that adding more team members will actually decrease work output.

Motivation losses occur when one or more group members disagree with the opinion of the majority and/or when subordinates disagree with the decisions made by team leaders or management.

Culture, Team Formation, Theories of Action, Defensiveness

Teams form by acquiring culture. Culture can be described as the set of values that govern team behavior. Some values are espoused, others are tacit and unacknowledged, but more deeply held. Per the theory-of-action model, espoused values govern behavior when nothing important is at stake, but deeply held values govern behavior in important situations.

When interfacing with people outside the team we tend to support our teammates even when they are wrong and we tend to support the team position even when we disagree with it. Model I behavior is accentuated by team membership. Teams contribute to defensiveness and inhibit learning.

Interfaces between Teams

The keys to effective management of interfaces between teams are:

1) Effective identification and sharing of objectives,
2) Effective sharing of data, where much of the data of interest is generated by the design efforts of the teams,
3) Effective communication between teams during synthesis and analysis as occasions arise which demand such communication, and
4) Effective testing of individual team designs against the overall project objectives.

Shared-mental-model Teams

A key point to take away from this chapter is the concept of shared-mental-model teams. Interfaces across team boundaries can be effective only if the interfacing parties have achieved this level of teamwork competence.

Key Factors for Effective Teamwork

We should expect the most effective teamwork when:

- The team is small.
- Team members agree strongly with the team goal and with the action plan to achieve it.

- There is a high level of trust between team members (minimizes Model I behavior).
- All team members are individually competent.
- All team members understand the role of all other team members (shared-mental-model level of team competence).
- Diversity on the team and respect for differing opinions will limit the tendency of the team to adopt extreme positions.

8.15 ANSWERING THE FOUR QUESTIONS

Why do so many changes occur late in project?

Poor early decision caused by groupthink

Teams will frequently yield better decisions than individuals because of a broader skill set, but this is not always the case. Team membership tends to make people more confident in their opinions because everyone in the group agrees. As a result, groupthink can drive a team to extreme positions, even more extreme than any individual member would take on his own.

Inability of a team to identify its own errors

Because teams drive members to commonality of thought it is difficult for team members to identify team errors and team members who disagree with the majority may not voice their opinion. Often errors made early in the project will not be detected until viewed by outsiders.

Inadequate testing of designs against all project objectives:

Individual teams usually will not completely understand the overall project objectives and the specific objectives of other teams. If the designs of individual teams are not tested against all of the project objectives it is likely that some aspects of the team's design will be inconsistent with another team's objectives.

Why do we have so many problems at interfaces?

Failure to share objectives

Failure to share data/information

Inadequate communication between teams during synthesis and analysis.

Inadequate testing of the design against the objectives of other teams.

Social pressure to support team

We are expected to support our teammates and to support the team position even when we disagree.

In-group favoritism causes lower opinion of out-groups

The very fact of belonging to one team can cause us to think less of members of another team. This lower opinion affects our interaction with the other team. We are less likely to value opinions of out-group team members than the opinions of in-group members.

Too many teams, too many interfaces

As projects become larger and more complex we have more teams and more highly specialized teams. The interfaces become more numerous and more difficult to manage.

Why do we repeat past mistakes?

Groupthink makes us overly confident of the correctness of our opinions

It is difficult to have much confidence in an opinion if no one else shares your view. Teams tend to share opinions; often there is coercion to believe what the team believes. Having several teammates share our opinions can make us over-confidence in our opinions and hence more resistant to learning.

Teams inhibit learning by increasing Model I behavior

Model I behavior is increased by team formation. We will often defend a team position even if we disagree with it.

Inability to learn from other teams we consider inferior

Our need for self-esteem causes us to favor our own team and to think that we are smarter, faster, more handsome, etc. than other teams. This bias makes it difficult for us to learn from other teams which we consider inferior, even when they are succeeding and we are not.

Status quo satisfies the objectives of one stakeholder

Not everyone shares the same objectives. A course of action that is judged to be wrong by some people may still be applied if it satisfies, or appears to satisfy, important objectives of one or more stakeholders. Example: Operations may object to a piece of equipment claiming that it doesn't work or is difficult to operate; but engineering may continue to specify it because it is small, cheap, easy to design, etc.

Why do we so often finish projects behind schedule?

Process losses caused by team coordination issues

Teams tend to slow us down. In a large team, if everyone takes a chance to talk and express/argue opinions, the decision making process can be much longer than if a single individual were empowered to make the decision.

Motivation loss caused by teams

Teams can energize individuals. They can also demoralize.

Chapter 9
Summary and Suggestions

Preface:

The objectives of the course laid out in Chapter 1 are:

1) To provide a general overview of important findings in the field of decision theory
2) To improve your decision making skills
3) To identify and recommend solutions to the common engineering problems.

The first 8 chapters were devoted to accomplishing item 1. This chapter focuses on items 2 and 3.

At the end of each chapter in the course we've tried to answer four questions:

- **Why do so many changes occur late in projects?**
- **Why do we have so many problems at interfaces?**
- **Why do we repeat past mistakes?**
- **Why do projects usually finish behind schedule?**

The first five sections of this chapter summarize the insights from each chapter and suggest solutions.

Following that insights from the course are used to develop a suggested decision strategy.

If the only tool you have is a hammer, everything looks like a nail.

CHAPTER 9 CONTENTS

9.1 WHY DO SO MANY CHANGES OCCUR LATE IN PROJECTS?
9.2 WHY DO WE HAVE SO MANY PROBLEMS AT INTERFACES?
9.3 WHY DO WE REPEAT PAST MISTAKES?
9.4 WHY DO WE USUALLY FINISH BEHIND SCHEDULE?
9.5 SUGGESTIONS SUMMARY
9.6 DECISION STRATEGY
9.7 DECISION STRATEGY - WHOSE DECISION IS THIS?
9.8 LITTLE AT STAKE – RANDOM SELECTION DECISION MAKING
9.9 SIMPLE OR TYPICAL – RULE-BASED DECISION MAKING
9.10 TIME PRESSURE – KNOWLEDGE-BASED DECISION MAKING
9.11 DECISION STRATEGY – ANALYTICAL DECISION MAKING
9.12 DECISION STRATEGY – CLOSING THOUGHTS
9.13 CLOSING THOUGHTS

9.1 WHY DO SO MANY CHANGES OCCUR LATE IN PROJECTS?

From Chapter 2 – Belief Generation and Sensemaking (See Section 2.22)

- Flawed beliefs caused by biases (availability error, illusion of validity) in data evaluation cause us to choose poor alternatives.
- Anchoring/Sensemaking – Once we embark on a course of action it is difficult to change direction.
- Risks underestimated due to familiarity (aspect heuristic)
- Frame of reference error (sunk cost).
- Design frozen too early making it difficult to correct errors.
- Team underestimates the risk because they like the design (affect heuristic)

From Chapter 3 – Objectives and Alternatives (See Section 3.17)

- Objective(s) identified late. We begin designing before we know all of the objectives.
- Objectives not fully defined or understood.
- Attractive alternative identified late.

From Chapter 4 – Deciding (See Section 4.29)

- Errors caused by ineffective decision strategies. In particular, decision made quickly via fast and frugal heuristics may be challenged later.
- Different stakeholders have different levels of risk tolerance. Designs considered acceptable by the design team may later be challenged as too risky.

From Chapter 5 – Acting (See Section 5.37)

- Planning errors discovered too late.
- Difficult to admit error (Model I behavior).
- Commitment – difficult to change plans after commitments have been made.
- Design substantially completed prior to sharing results with other teams. Anchoring on design makes it difficult to change.

From Chapter 6 – Learning (See Section 6.19)

- Defensiveness makes learning difficult (Model I behavior). Learning is often delayed until someone not involved in the design flags a problem publicly.

From Chapter 7 – Stress and Expertise (See Section 7.20)

- Inadequate mental simulation of the design. Failure to identify things that might go wrong.
- Focus on plausible objectives. We have a tendency to focus on objectives that are easily achieved and discount those that are more difficult.

From Chapter 8 – Teams and Interfaces (See Section 8.15)

- Poor early decision caused by groupthink.
- Inability of a team to identify its own errors due to lack of diversity on the team.
- Inadequate testing of designs against other teams objectives.

Let's put these observations into the context of the engineering model:

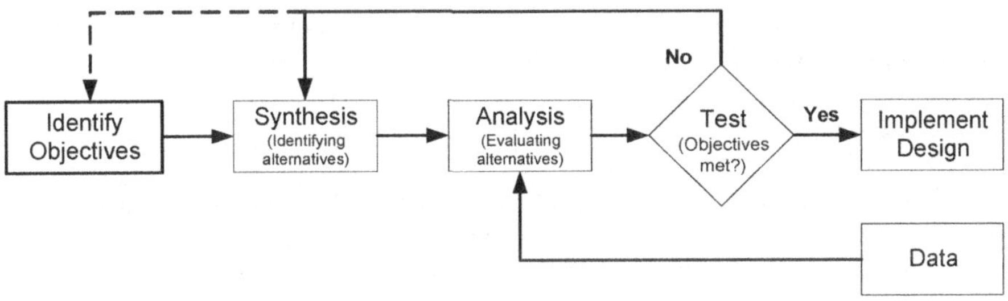

Objectives

Failure to effectively identify design objectives early in the process will clearly lead to problems down the road. Indeed, I believe that that is the single largest cause of late changes. There are several reasons why we fail to identify objectives adequately. We've discussed these more than once already, but I want to focus attention again on five of them:

- Alternative-focused thinking – our natural desire to try to accomplish something quickly.
- Interface issues – difficulty in identifying the objectives of other stakeholders.
- Our tendency to define problems that we know how to solve – the objectives are not givens, we choose the objectives and we will tend to choose objectives that are achievable so as to guarantee success.
- We tend to design for steady state operation in as-built condition and hence to miss objectives related to construction, commissioning, maintenance, and non-steady state operations such as startups, shutdowns, upsets.
- We tend to focus on readily measurable objectives (such as cost).
- We underestimate risk inherent in a preferred option (affect heuristic).

Synthesis, Analysis and Testing

The next three steps in the engineering process involve identifying an alternative, evaluating the alternative and the testing whether that alternative adequately satisfies the objectives. Many things can go wrong in this process. We may fail to identify an excellent alternative through failure of imagination, or lack of effort. Our analysis of a proposed design may be flawed due to incompetence or flawed data. But I want to focus mainly on the testing step where we ask "Does this alternative adequately satisfy all of the design objectives?" This is a fuzzy question. What does "adequate" mean? Different people will have different opinions. The testing step can go wrong for a number of reasons including:

- If you don't fully understand the objectives you cannot adequately judge whether an alternative satisfies them. This judgment can be particularly difficult when other teams and other stakeholders are involved. How do you judge whether your design meets other people's objectives? Will they be satisfied with your decision?
- We only evaluate the options that we consider attractive. An attractive option will likely be one that effectively satisfies our most important objective. We can easily get anchored on that alternative and our evaluation will be biased to show that it adequately satisfies the other objectives as well.
- Tunnel vision; under time pressure we tend to focus on some objectives to the exclusion of others.
- Risks underestimated: A cohesive team will likely have a high opinion of the team design and will naturally overestimate the positive aspects of the design and underestimate the risks.
- There is a tendency for teams to substantially complete their design prior to sharing results with other teams.

Iteration / Learning

The engineering process is iterative. Our first ideas usually have some flaws. We identify problems and cycle back to identify alternatives to address those problems. This cycle is a learning loop. Each time around the loop we learn some things that work or some things that don't or some of both. Each loop may also help identify objectives and alternatives initially missed.

> *The more often you make this loop and the more effectively you learn on each loop, the better your design will be.*

This observation is analogous to the OODA loop (Observe, Orient, Decide, Act) observation of Colonel John Boyd which changed the way the U.S. Marines fight battles (See *Boyd*, by Robert Coram).

Making this learning loop effectively and quickly is a key to effective engineering design. What are the factors in making the loop effectively?

- Shared objectives and values so that the person making the evaluations will make them confidently. Not knowing what other people will think about a particular decision will slow decision making.
- Trust: This is related to the first bullet. You can quickly and effectively make decisions only if you trust that others will agree with your decision and back you up.
- Moral conviction: You can make decisions more confidently when you believe that you are doing the right thing for the right reasons. Conversely large motivational losses occur when people are expected to do unethical things.
- Appropriate analysis tools. The slow part of the loop is frequently the analysis step. A key to executing the learning loop quickly is conducting the analysis quickly. But many of our tools are designed for rigor and accuracy rather than speed. Often these tools can only be run by experts so there is a large time

delay in getting results. In an environment with significant uncertainty we need simpler tools that are fast and accurate enough (plausible).

Correcting Errors in Time

Eventually you accept the design and move into implementation (detailed design, purchasing, construction, operation). This occurs either when you find an option that fulfills all of your objectives adequately or you run out of time or engineering budget.

Once you have chosen a design for implementation the entire dynamic of the effort changes. Project teams often know they are in trouble, but take little action to correct mistakes. They anchor on the selected alternative and energetically dismiss challenges to the design. Admitting errors would cause them embarrassment and might cause them to lose control and hence does not happen easily (Model I behavior). Project audits and safety reviews (such as HAZOPs) often do not have the impact that they should because the design team becomes defensive and argues away problems rather than implementing the recommended solutions. Double-loop learning is required in order to voluntarily make changes; you will have to admit that your original design was flawed. Changes in this phase may occur only when they are forced upon the design team by outsiders via audits and reviews.

Resistance to change is both necessary and problematic. It is necessary because you have to get on with the project (avoid unnecessary changes); but it is problematic when it prevents you from changing flawed decisions.

SUGGESTED SOLUTIONS -- FREQUENT LATE CHANGES:

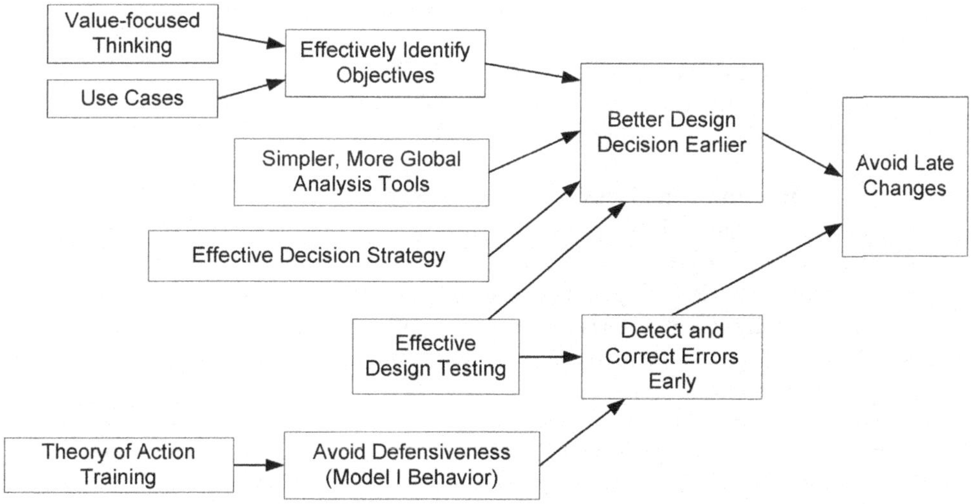

The graphic above is a means objectives network with the fundamental objective of avoiding late changes. Some of the objectives above are discussed in section 9.5.

9.2 WHY DO WE HAVE SO MANY PROBLEMS AT INTERFACES?

Chapter 2 - Belief Generation and Sensemaking (See Section 2.22)
- Failure to share data due to structural secrecy
- Data means different things to different people.
- Social pressure to support team.

Chapter 3 – Objectives and Alternatives (See Section 3.17)
- Objectives not fully defined or understood.
- Different teams develop incompatible solutions.
- Different teams disagree on the best alternative.
- Teams have conflicting objectives.

Chapter 4 – Deciding (See Section 4.29)
- Different strategy used for selling decision across interfaces than for making decisions within the team.

Chapter 5 – Acting (See Section 5.37)
- Model I behavior (defensive behavior is accentuated by team membership).
- Ambiguity in plans.
- Responsibility ambiguity. If it's not clear whose responsibility a task is it may not get done.
- Design substantially completed prior to sharing results with other teams. Anchoring on design makes it difficult to change.

Chapter 6 – Learning (See Section 6.19)
- When there is conflict it is relatively easy to see error in the other team's approach and difficult to see errors in your own or your team's work.

Chapter 7 – Stress and Expertise (See Section 7.20)
- Decisions made easily by one team may be difficult to sell to another team with less expertise in the subject.

Chapter 8 – Teams and Interfaces (See Section 8.15)
- Failure to share objectives.
- Failure to share data/information.
- Inadequate testing of designs against other team's objectives.
- In-group favoritism causes lower opinion of out-groups.
- Too many teams, too many interfaces.
- Social pressure to support the team position.

Summary - Keys to Effective Interfaces

We've already discussed the keys to effective interface management in Chapter 8:
- Effective identification and sharing of objectives.
- Effective sharing of data, where much of the data of interest is generated by the design efforts of the teams.
- Effective communication between teams during synthesis and analysis as occasions arise which demand such communication.
- Effective testing of individual team designs against overall project objectives.

There are some reasons why this is particularly difficult across team interfaces including:

- Model I behavior (defensiveness) is accentuated by team membership.
- In-group favoritism causes us to have a lower opinion of other teams and to more easily dismiss their objectives and opinions.
- Different skill sets makes communication with other teams more difficult than communication within a team.
- Teams often have competing objectives.

SUGGESTED SOLUTIONS - PROBLEMS AT INTERFACE

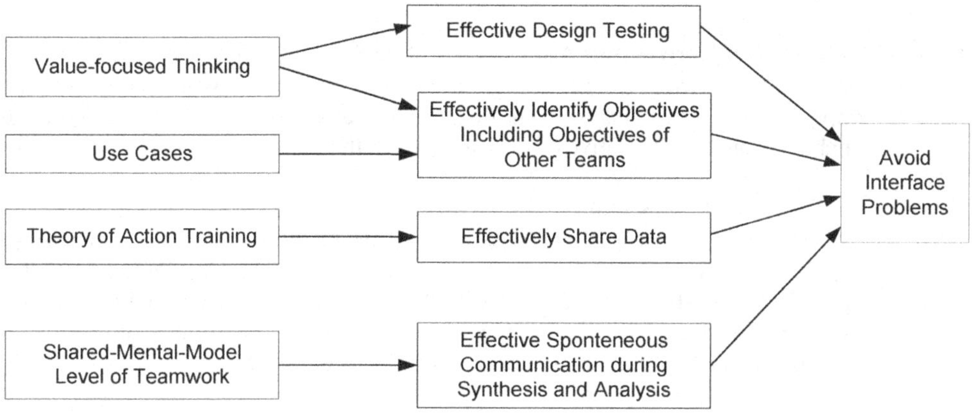

9.3 WHY DO WE REPEAT PAST MISTAKES?

From Chapter 2 – Belief Generation and Sensemaking (See Section 2.22)

- Anchoring on past success: A practice that has worked in the past is difficult to abandon.
- Sensemaking: History is more complicated than history books suggest. It is usually possible to craft more than one story to explain a poor result. It is easy to "learn" the wrong lessons.
- Salience: It is difficult to challenge practices that seem obviously correct.
- Structural Secrecy: Managers who could affect meaningful change on the next project are shielded from the true causes of error by the organizational structure.

From Chapter 3 – Objectives and Alternatives (See Section 3.17)

- Alternative-focused thinking: A focus on objectives would give us the opportunity to find better solutions. Alternative-focused thinking tends to anchor us on the alternatives we've applied in the past.

Chapter 4 – Deciding (See Section 4.29)

- Frugal decision models make learning difficult and unreliable. When we apply a detailed analytical effort to make a hard decision, we have a basis for investigating what when wrong with the analysis. When we make decisions via fast and frugal heuristics there may be little information in the decision model to guide us in making a better decision the next time.

From Chapter 5 – Acting (See Section 5.37)

- Inability to recognize unactionable plans: A great deal of business advice recommends Model II behavior as a solution to problems created by Model I behavior. Since people are incapable of sustained Model II behavior the advice is not actionable.
- Systems effects complicate learning: Actions often have unintended side effects that undo the good that the actions were supposed to accomplish. It is difficult to recognize those side effects, especially if there is a time delay between the apparently positive outcomes which occur quickly and the negative effects which occur later.
- If commitments have been made then changing course can be difficult.

From Chapter 6 – Learning (See Section 6.19)

- Simply recognizing a problem is not learning.
- Failure to learn difficult lessons (double-loop learning) due to Model I behavior.
- Many mistakes have multiple causes: The obvious cause of a problem may not be the only cause or the most important.
- Sensemaking (making sense of an event rather than making corrections).
- Lack of motivation to change: Real learning occurs when we take action to solve problems which are important to us. We may continue in a failed course of action if the results are not that important to us or if some side effects are more important.
- The fact that one or more members of an organization learn something doesn't mean that the organization knows it.

From Chapter 7 – Stress and Expertise (See Section 7.20)

- It may be more difficult to recognize errors in decisions made naturalistically: Consider a case where you actively consider different alternatives before selecting one that didn't work. In this case you have ready fodder for learning. It is easy to ask "what if I had chosen option __ instead?" In the case of a decision made naturalistically you may not have considered any other options and hence will not have ready access to a feasible solution to compare to the one you chose.

From Chapter 8 – Teams and Interfaces (See Section 8.15)

- Groupthink makes us overly confident of the correctness of our opinions.
- Teams inhibit learning by increasing Model I behavior. We will often defend a team position even if we disagree with it.
- Inability to learn from other teams we consider inferior: Our self-esteem causes us to favor our own team and to think that we are smarter, faster, more handsome, etc. than other teams. This bias makes it difficult for us to learn from other teams which we consider inferior, even when they are succeeding and we are not.
- We may persist in errors when the status quo satisfies the objectives of one stakeholder with the power to influence the decision.

Summary

A rich story emerges from these observations.

It is difficult to know the whole story when a project goes bad. Projects are complex undertakings. Actions have their intended effects and unintended effects (systems effects). Different people will develop different views of what went wrong and why it went wrong. Each of these views will be a simplification of what actually happened (sensemaking). These stories may not be rich enough, correct enough, or consistent enough to provide a consensus view on how to approach the problem on the next project.

Salience may be the most important factor is evaluating a course of action. If a plan seems obviously correct it is very difficult to learn that it is wrong.

When the next project comes along the memories of the last one will have faded some. Rather than focus carefully on all of our objectives including how/why some of them were not achieved on the last project we fall into what comes naturally – alternative-focused thinking. Since the alternatives we've applied in the past are most easily identified it is likely we will repeat past actions.

Project teams often know they are in trouble, but take little action to correct mistakes. Admitting errors would cause them embarrassment and might cause them to lose control and hence does not happen easily (Model I behavior). The same face-saving, defensive behavior sabotages post-mortem lessons learned studies so that the real causes of error remain hidden. The structure of organizations contributes to this. When a project is in trouble the worker bees often recognize problems that project management doesn't see (structural secrecy).

When we recognize a problem we are strongly biased to "learn" the least threatening lesson (Model I behavior, single-loop learning). The more important a lesson is, the harder it is to learn. Much of what we do "learn" is unactionable. Many project problems are caused by Model I behavior. The obvious solution is to get people to behave better. But people are not consistently capable of open, trusting Model II behavior in threatening or important situations.

There is a real paradox in learning from others. We should use industry experience to learn from other peoples mistakes. Arguably it should be easier for us to learn from other people's mistakes because such learning is non-threatening. But this is countered by in-group favoritism. If we view the other team/company that made the error as inferior to us we may believe that we have little to learn from their mistakes. Another problem with learning from other people's mistakes is that it is very difficult to get enough information about what actually went wrong.

We may repeat a course of action even when most people consider the course to be wrong if that course satisfies the objectives of one or more stakeholders who have influence. The Accounting and Purchasing groups may develop contracting strategies that make no sense to engineering. Engineering will design systems and select equipment that Operations opposes. These actions may be intentional power plays, but are probably often done out of ignorance. We never know all the objectives of other stakeholders (structural secrecy) and hence may continuously do things they would consider errors.

Finally, mistakes are often made because the organization does not know all that individuals in the organization know. Some knowledge is not readily available to the right people at the right time. Lessons learned by individuals do not translate into lessons learned by the organization unless effective Organizational Learning Mechanisms are in place.

SUGGESTED SOLUTIONS – TO AVOID REPEATING MISTAKES

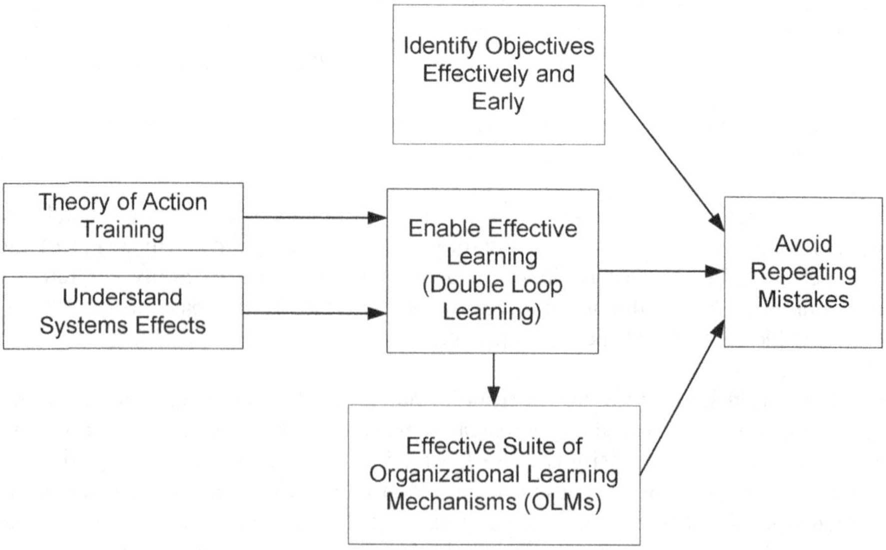

9.4 WHY DO WE USUALLY FINISH BEHIND SCHEDULE?

From Chapter 2 – Belief Generation and Sensemaking (See Section 2.22)

- Sensemaking of the future: We simplify the future in the same way that we simplify the past. Sensemaking of the future results in overly simplistic and optimistic schedule projections.
- Structural Secrecy: Schedules are important to project management. Schedules are often optimistic because management puts pressure on subordinates to make optimistic projections. As problems occur subordinates are often aware of schedule problems long before project management.
- Collecting too much data.
- Data means different things to different people: As we attempt to make decisions with multiple stakeholders we have difficulty when different people interpret data in different ways.
- Dissonance reduction is time consuming: Making a difficult decision is time consuming, and the hard work isn't even finished with the decision. Hard decisions create anxiety. Anxiety reduction takes time and action may be impeded during this period.

From Chapter 3 – Objectives and Alternatives (See Section 3.17)

- Trying to satisfy unrealistic objectives: Engineers writing specifications tend to ask for the moon without thinking through the consequences. Other engineers trying to achieve the specification then spend inordinate amounts of time trying to achieve the impossible.
- Evaluating too many alternatives: Carrying too many alternatives for too long costs time and money and distracts attention from the real issues.

From Chapter 4 – Deciding (See Section 4.29)

- Applying inappropriate decision strategies: We waste time when we apply too much effort to simple problems.
- Multiple teams and multiple contractors all need to finish their parts on time for the project to finish on time.

From Chapter 5 – Acting (See Section 5.37)

- Unrealistic schedules caused by the planning fallacy: Early plans tend to be high level and incomplete and hence optimistic. Also, we tend to plan for success. But assuming that everything will go right is not realistic.
- Inadequate plans, especially omission of required tasks.
- Unactionable plans. Plans that some stakeholders disagree with may be difficult to implement.

From Chapter 6 – Learning (See Section 6.19)

- Failure to learn from missed schedules: Since we routinely miss schedules there are multiple opportunities for learning the reasons why, yet we do not. This is a particular instance of repeating past mistakes.
- Learning the wrong lessons from missed schedules (sensemaking): The history of past projects is not unambiguous. We never know all the factors that went into making a project successful or unsuccessful. It is easy, given a partial history, to craft an explanation for a past error that is wrong or incomplete. It is easy to attribute failure to bad luck.

From Chapter 8 – Teams and Interfaces (See Section 8.15)

- Process losses caused by team coordination issues: Teams tend to slow us down. The larger the team, the greater the coordination losses.
- Motivation loss caused by teams: Teams can energize individuals. They can also demoralize.

SUMMARY

There are three main reasons we miss schedules:

- Optimistic scheduling.
- Failure to correct the schedule when problems occur.
- Slow execution.

Optimistic scheduling

Sensemaking and the Planning fallacy

History is much more complicated that the history books suggest. We make sense of history by simplifying it. We tend to view the past as much simpler and more deterministic than it really was. When an event occurs we can read past actions as leading inevitably to that result even if it would have been difficult to make those predictions before the fact.

We make the same errors in predicting the future. Our vision of the future is simpler than the future really will be. The actions we take will not lead inevitably to our desired result. The future may unfold in countless different ways that could impact the project. When we make sense of the future in this way we naturally view our tasks as simpler to accomplish than they really are.

Plan-based scheduling is biased towards assuming success at each step. But it is not realistic to think that everything will go as planned. One problem with accounting for inevitable failures in our plans is that we cannot predict a priori which plans will fail or take longer than scheduled.

Management Pressure

Management is very interested in the project schedule and will often exert pressure on the design team to propose an optimistic schedule. As a result the design team is often saddled with a schedule that they don't think they can meet. This can be very demoralizing causing poor project execution.

Early Enthusiasm

Early plans tend to be ambitions because they are generated early in the project when enthusiasm is high.

In-Group Favoritism

We are biased to believe that we are better than others; better than other teams, better than other companies. It is common for anyone who belongs to a team to think that his/her team is better than other teams. It is not hard to believe that **we** can perform projects faster than **they** can.

Failure to correct schedule when problems occur

If early schedules are optimistic then it is likely that we will experience problems. But problems on the project which should result in schedule changes often do not result in formal schedule changes. It is typical to stick to the original schedule long past the point where that schedule becomes unrealistic. Perhaps the greatest risk of seriously busting the schedule comes from valiant attempts to meet an unrealistic schedule. Foolish decisions made in trying to achieve unrealistic schedules can cause serious schedule delays. There are several reasons for failure to correct schedules:

<u>Anchoring:</u>

The original schedule is an anchor point. We tend not to adjust it or not to adjust it enough.

<u>Defensiveness (Model I behavior):</u>

Admitting that we are running late may be embarrassing and so this information may be covered up. This may be especially true of vendors and outside contractors who may cover up and work to get back on schedule rather than admit being behind.

<u>Structural Secrecy</u>

The first people to realize schedule problems are likely to be subordinates. The story may not find its way to project management, may take some time to get there, and may be very watered down when it arrives. Managers cannot know the details that their subordinates know. If project management doesn't want to hear about schedule problems, then they probably won't.

Ineffective execution

Ineffective execution can cause schedule delays. Typical reasons include:

<u>Coordination losses caused by teams and interfaces between teams</u>

When multiple people work on related tasks their activities need to be coordinated. Coordination losses tend to increase exponentially with the size of the team. For large teams the effort required to share data and coordinate actions and decisions can surpass the amount of time spent on productive labor.

<u>Motivation losses caused by teams</u>

We all feel the need to be in control especially when something important is at stake. When we disagree with management direction or the opinion of the team, this loss of control is demotivating. This is particularly true on teams managed by compulsive micro-managers.

<u>Decision take too long because of ineffective decision strategies</u>

The vast majority of engineering decisions are made quickly via simple heuristics (engineering judgment). But we often overwork relatively simple issues. This is particularly true of issues that attract management's attention. Some decisions are easy to make, but difficult to sell to management and/or to other teams. Often more work goes into defending a decision than into making it.

Values not agreed to and shared

Decision making is facilitated by shared values. If you know what others on the project will think about a particular issue then you can confidently make the decision. Conversely, if you don't know what other's values are and you don't know what decision they would make or support, then decision making is dramatically slowed. Decisions that cross team boundaries can be particularly time consuming and difficult to make.

Correcting errors / making changes

Making changes and correcting errors causes rework, coordination losses and further errors. Unnecessary changes are very time consuming, hence projects seek to minimize changes. The downside to minimizing changes is that anchoring on early decisions makes it difficult to correct errors. Because of the reluctance to make changes, many early errors linger until late in the project when changing them is the most time consuming.

Setting unrealistic objectives

It is common for design teams to "shoot for the stars." This is especially true when we write specifications that other people will implement. In such situations "gold plating" can seem free. But the engineering and testing effort and potential failure to meet such specifications can be very time consuming.

Multiple Teams, Multiple Contractors

Multiple teams and multiple contractors are involved in most project organizations. Each team has to finish on time in order for the project to finish of time. The more teams there are, the more likely it is that at least one will finish late.

SUGGESTED SOLUTIONS – SCHEDULE FAILURE

Addressing schedule failure is more challenging than addressing the other three questions. There are so many factors that impact schedule failure that it is not entirely clear how to proceed. The graphic below is probably a good start. Late changes cause schedule problems. Interface problems cause schedule problems. Failure to learn from past mistakes causes schedule problems. Avoiding schedule problems requires fixing all of those problems.

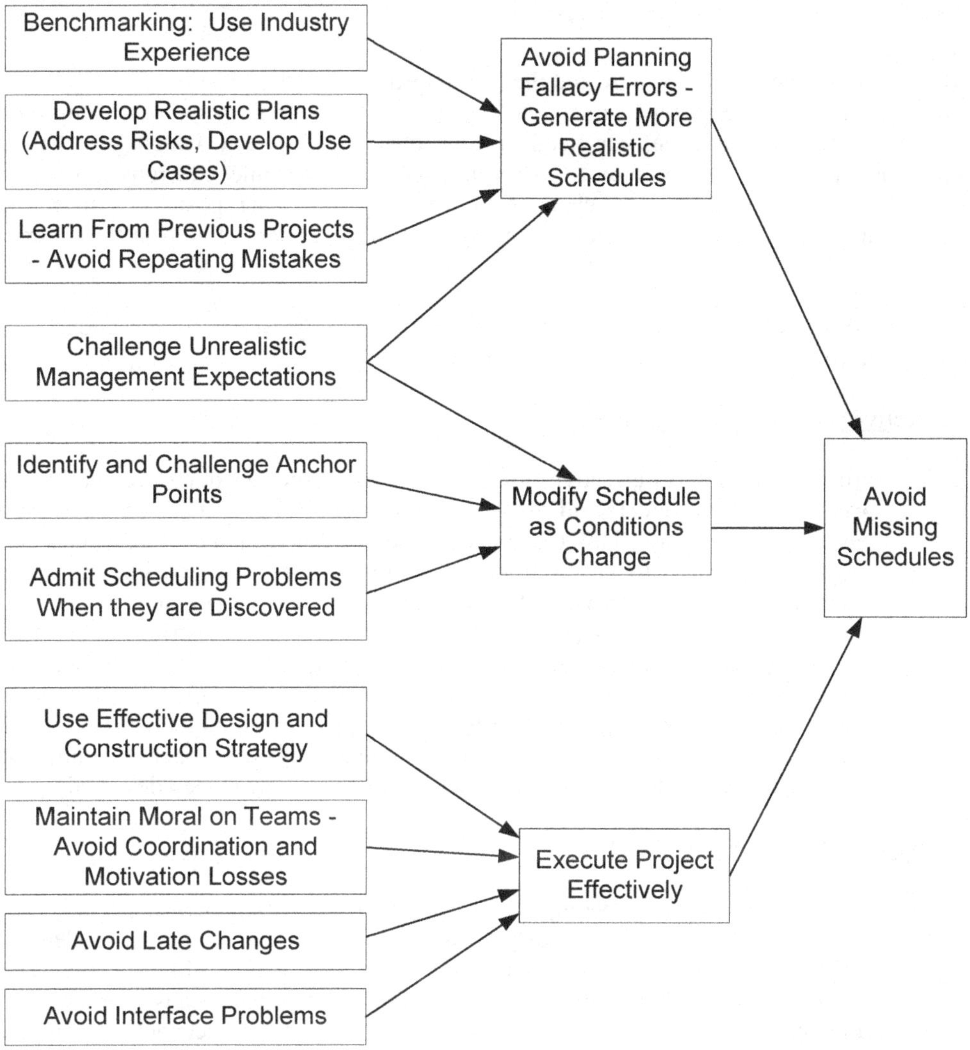
Benchmarking: Use Industry Experience
Develop Realistic Plans (Address Risks, Develop Use Cases)
Learn From Previous Projects - Avoid Repeating Mistakes
Challenge Unrealistic Management Expectations
Identify and Challenge Anchor Points
Admit Scheduling Problems When they are Discovered
Use Effective Design and Construction Strategy
Maintain Moral on Teams - Avoid Coordination and Motivation Losses
Avoid Late Changes
Avoid Interface Problems
Avoid Planning Fallacy Errors - Generate More Realistic Schedules
Modify Schedule as Conditions Change
Execute Project Effectively
Avoid Missing Schedules

9.5 SUGGESTIONS SUMMARY

Until the last section we have treated the four questions as independent entities. They are in fact related as the graphic in section 9.4 shows. If you want to work on improving your schedule performance, you need to understand the degree to which late changes, interface problems and failure to learn impact your schedule overruns. Likewise interface problems cause late changes and vice versa. The positive spin to that observation is that any improvement you make in one area should help in the others as well.

There are several recurrent and important themes in the suggestions from sections 9.1 to 9.4. I think the most important ones are these:

Effectively Identify the Objectives

> Probably the key to effective design is setting the design objectives early. We want all teams to understand the overall project objectives. Further, we want all teams to understand all of the objectives of the each of the other teams. My suggested approach to this is to conduct structured exercises to identify project objectives following the value-focused thinking approach outlined in chapter 2 and to develop use cases (section 3.6). In the process industries a use case is effectively an operating procedure. It is fairly common practice to write the operating procedures after a process has been designed. Operating procedures should be written at a high level early in the design effort and matured along with the project. We need to have engineers designing things to be operated rather than trying to determine how to operate processes after they have been designed.

Develop Effective Action Plans

> Action plans are often incomplete, optimistic and unactionable. Especially when action plans are developed early in a project they tend to be incomplete and optimistic (see discussion of the planning fallacy in section 5.8). Use the Means Objectives Network structure (Chapter 2) to build complete plans or to check for completeness. It's not enough to have an accurate and complete action plan. You need a plan which will actually be implemented correctly. The discussion on Human Error (sections 5.7, and 5.9 to 5.12) provides guidelines for avoiding error in the implementation of action plans. Unactionable action plans can result when one or more stakeholders disagree with the plan. Stakeholders who disagree with the plan may doom an otherwise brilliant plan. For insights into this behavior see the discussion of Theories of Action (sections 5.14 to 5.26).

Simpler, More Global Analysis Tools

> It's my opinion that we have made engineering overly complex. In oil development projects Reservoir Engineers play with sophisticated models to estimate oil production over time from a reservoir. These tools are expensive and only a few people in the company can run them. Pipeline Engineers, Subsea Engineers, Topsides Process Engineers all have their own sophisticated tools. Only a handful of people in the company run any one of these tools and no one runs them all. It is difficult for anyone to see the big picture. Some tools integrate other sophisticated tools together, but they are even more expensive and even fewer people have access. Further, rather than using these tools to clarify the design environment and develop project design basis data, they are often play the

opposite role. These tools are used extensively to study sensitivities to the point where the design targets become more and more unclear over time.

The fallacy in this approach is that the reservoir properties are probably never known to within an accuracy of better than $\pm$ 20%. It makes little sense to spend so much effort running rigorous programs in this environment. We need simpler programs that conform to the actual design environment and which more people can understand and use.

Develop Shared-Mental-Model Level of Team Competence

We discussed four levels of team competence in section 8.4;

1) Individual Team Member Competence,
2) Shared Goals,
3) Shared Mental Models, and
4) Shared Awareness.

Truly effective teamwork occurs at the shared-mental-model level. Section 8.5 provides suggestions for developing shared-mental-model teams.

Identify and Correct Design Errors Early via Design Testing

Engineering is an iterative game. Our first ideas rarely satisfy all of our design objectives, yet we frequently anchor on early ideas because they effectively accomplish our key objectives. Effective and efficient engineering depends on effectively identifying and correcting design errors as early as possible. The first step in this has to be effective identification of the design objectives. These objectives must be made explicit and must be actively used by all project members. In particular, the objectives have to be remembered and referred to when changes are made. The design must be tested at each phase in the design and whenever changes are made to ensure that the identified objectives are achieved.

This advice is admittedly vague. Exactly what sort of testing is appropriate will depend on the design. Measurable objectives assist in developing appropriate tests. When equipment is purchased the purchase specification should detail the testing to be performed prior to acceptance of the equipment.

Dynamic simulators now provide a very effective test vehicle for process designs and process control systems.

Effective commissioning procedures (tests) greatly simplify startup and early life operation by catching design problems and manufacturing and construction errors.

Apply an Effective Decision Strategy

The last few sections of the course are devoted to development of an effective decision strategy. Effective engineering depends on effective decision making. Simple decisions should be made simply, the right level of effort should be devoted to difficult decisions, and we should be able to tell the difference between simple and difficult decisions.

9.6 DECISION STRATEGY

To close out the course the next few sections present a suggested decision strategy.

For general decision making consider the following questions in this order:

- **Whose decision is this to make and who are the stakeholders?**
- **How much is at stake?**
- **How complex or unusual is the decision problem?**
- **How much time pressure are you under?**
- **What are the important features of the decision problem?**

"Whose decision?" is a good question to ask first because the identity of the decision maker(s) and the stakeholder(s) is critical to many decision problems. Also, this question has to be asked before you ask how much is at stake because that question may have a different answer for different stakeholders.

"How much is at stake?" is a key early question. It helps you avoid wasting a lot of time and effort on unimportant problems.

"How complex or unusual?" is the third question. If the problem is simple or typical it should be easily solved even if a great deal is at stake.

"How much time pressure?." This question only occurs if the stakes are high and the problem is complex or unusual. People don't feel stress if nothing is at stake and/or if the problem is simple.

The final question "What are the important features?" refers mainly to the features of an analytical decision process. You will only get this far when the stakes are high, the problem is non-trivial and time is available to analyze the problem.

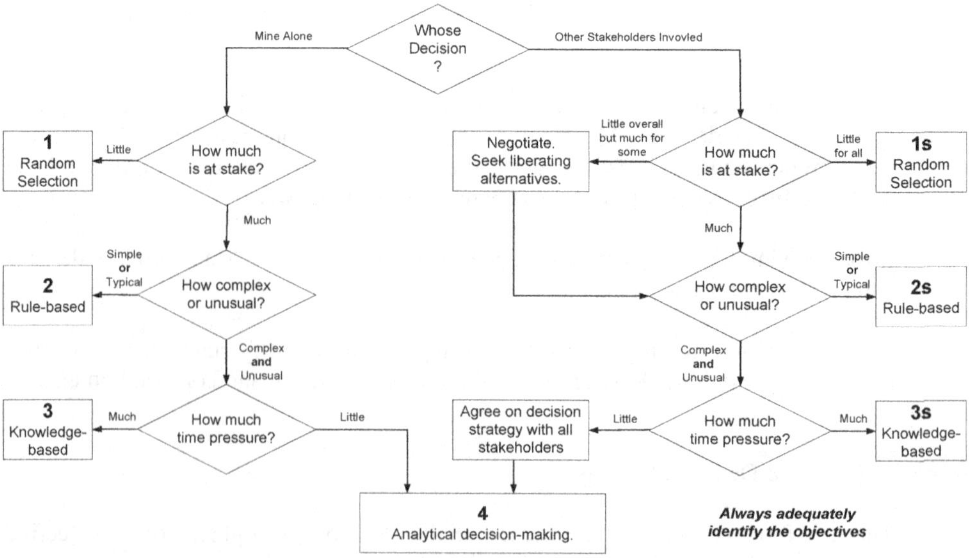

The answer to the first question divides the map into two roughly symmetrical halves. On the left are individual decisions, on the right are group decisions.

Nodes 1 and 1s – Random Selection Decisions

When the answer to the second question "How much is at stake?" is "Little" we have a problem that could effectively be solved by flipping a coin or using some other random selection method. If little is at stake it will make little difference which option you select.

When multiple stakeholders are involved the stakes may be different for different stakeholders. There may be little at stake for the project overall, but a great deal at stake for one or more individual stakeholders.

Nodes 2 and 2s – Rule-based Decisions

If you've made it this far then something significant is at stake, but the problem is simple or typical. If the problem is typical there should be a rule or a traditional solution available. If the problem is simple the solution should be readily apparent.

Nodes 3 and 3s – Knowledge-based Decisions

These are complex and unusual problems with little time available. Effective solution requires expert knowledge.

Node 4 – Analytical Decision Making

This is the type of problem that people typically think about when they think about decision making; much at stake, complex or unusual problem, and enough time available to work the issues.

9.7 DECISION STRATEGY - WHOSE DECISION IS THIS?

Your Decision

Decision making is at its simplest when the decision is yours to make and yours alone and you will not be required to defend or justify it.

Group Decision and/or Multiple Stakeholders:

Most decisions of interest involve other stakeholders and/or other levels of management. If you have to defend or sell your decision it is likely you'll have to work harder, collect more data, build a better model, etc. The temptation in such situations is to make the decision quickly (via naturalistic methods or fast, frugal heuristics) and to defend it via analytical techniques. Your decision is likely to be biased in favor of those aspects of the problem that are most important to you and/or your team. You are likely to spend more time justifying your decision than making it.

In this situation:

- Identify the stakeholders and decision maker(s).
- Try to get all stakeholders to agree on a decision strategy prior to making the decision.
- Consciously spend time identifying the objectives of all stakeholders. Focus initially on objectives rather than alternatives (value-focused thinking).
- Try to be objective in evaluating the alternatives against the objectives (value other stakeholders objectives).
- Think about the context. Would a better solution be available to someone higher in the organization? Should you escalate the decision up the organization?
- Or should the choice be left to the person implementing the solution?

Always be aware of turf issues (Model I behavior). Your decision may not be actionable if other stakeholders will lose even a small measure of control or influence?

Someone Else's Problem?

For both of the answers above we've assumed that you are responsible for the decision at least in some way and would continue working it. But the answer may be that this is someone else's responsibility entirely and should be passed on.

9.8 LITTLE AT STAKE – RANDOM SELECTION DECISION MAKING

The measure of what is at stake can be ambiguous. The appropriate measure should be the difference in results obtained from selecting the best option vs. selecting the second best.

Some decisions are inherently trivial – choosing the color of a car for instance. Others are potentially significant, but may be trivial. For instance, even if a multi-million dollar contract is on the line, there may be little at stake if there is no material difference between the top two bids.

Uncertainty complicates the determination of what is at stake especially if you are comparing a certain outcome against a risky option.

Consider several factors in making the judgment of what is at stake:

- What is the likely difference between results of the two best options?
- What is the percentage difference? Small differences in big numbers may have little validity.
- How much variability is there in the estimates? Do the uncertainties compare? Are you comparing a certain outcome against a risky alternative? How does the magnitude of the difference compare to your risk tolerance? For an option with significant uncertainty, what if the worst possible outcome occurs? What if the best occurs?

<u>Management Involvement</u>

Management involvement is likely to be a function of the total spend rather than the difference between alternatives.

<u>Identify your real objectives; It's not only money</u>

Be careful not to focus too much on the monetary impact of a decision. There may be other very important things at stake. Consider other objectives such as:

- Safety impact
- Environmental impact
- Schedule impact
- Reputation
- Ease of operation
- Etc.

1

Your Decision, Little at Stake

This is the simplest of all decision problems that surface into consciousness.

It is your decision to make; there are no other stakeholders. You won't need permission to implement it and you won't have to defend the decision to anyone and there is very little at stake.

Avoid Zone of Indifference Errors

Paradoxically, the less there is at stake, the more difficult a decision can be. *Avoid zone of indifference paralysis.* Don't spend your time on issues that don't matter. If there is no obvious advantage either way, then ***flip a coin and move on***.

1s

Stakeholders Involved, Little at Stake

These decisions should be as easy as a Node 1 decision; there is little at stake and a coin flip should suffice for making the decision. But other people are involved and so it may be much more difficult. Still, your objective has to be *avoiding zone of indifference paralysis.*

Point out to all stakeholders how little is at stake emphasizing that spending time on this decision is not worth the effort.

Different Stakes for Different Stakeholders

Even when very little is at stake overall, different alternatives may have different implications for different stakeholders. In this scenario:

Focus on the Objectives: Evaluate each stakeholder's objectives. It is likely that there will be significant agreement on overall objectives even if they disagree on the best alternative. Stakeholders who become committed to a particular alternative will naturally focus on the objectives that that alternative satisfies. By obtaining explicit agreement on the important objectives you may be able to demonstrate that there is less at stake than was initially thought. Focusing on objectives may also surface hidden agendas – or at least to make it harder to defend preferences based on hidden agendas.

Identify Liberating Alternatives and Negotiate: Evaluate the alternatives from each stakeholder's point of view in addition to taking a "best for project" view. Is another alternative only a little worse overall, but significantly more balanced? Can you identify a new alternative or combination of alternatives which is better for all? Can the "best for project" alternative be modified to make it more acceptable to some stakeholders? Can you trade a better outcome for the reluctant stakeholder on another issue in exchange for the "best for project" solution to this one?

Involve management if the problem cannot be solved at your level.

Disagreements over uncertainty and risk:

Honest disagreements often occur related to uncertainty estimation and risk assessment. This is common in safety and environmental decisions. (How safe is safe enough? How much money is an extra bit of safety worth?) It is easy to dismiss another stakeholder's fears of a highly unlikely outcome if that outcome has little impact on you.

9.9 SIMPLE OR TYPICAL PROBLEM – RULE-BASED DECISION MAKING

How Complex or Unusual?

Many of the problems we face on a daily basis are relatively simple, often because we've faced similar problems before. If the problem is simple or typical there is likely to be a simple or well accepted solution.

As we have seen in chapter 4 even apparently complex problems frequently yield to simple solution models. The complexity of the problem should be reflected in the complexity of the decision strategy: Ask:

- How complex is the environment?
- How well do you understand the environmental complexity? How "true" is your model?
- Can you collect adequate data to effectively apply a complex model?

What level of rigor does the available data justify? Are you spending a lot of time generating a $\pm$ 1% analysis based on $\pm$ 10% data?

Are you using the right tools? We tend to apply analytical tools to engineering problems simply because we can. Twenty years ago we ran process simulation programs once or a few times on a project to yield design data which we shared with others. Now we run simulations programs thousands of times – and it is often unclear what the valid design basis data is.

2

Your Decision, Much at Stake, Simple or Typical Problem

If the problem is typical there should be an established rule or a traditional, well accepted solution. **Apply the rule.**

In the absence of a rule, decide simply and quickly. Apply a decision model that matches the complexity of the problem.

2s

Stakeholders, Much at Stake, Simple or Typical Problem

Paradoxically, the existence of stakeholders in this situation may have less impact that in the Random Selection node (1s). If an established rule or tradition exists, **apply the rule**.

In the absence of a rule, determine an effective solution simply and quickly and propose the solution to the stakeholders. Apply a decision model that matches the complexity of the problem.

9.10 TIME PRESSURE – KNOWLEDGE-BASED DECISION MAKING

How Much Time Pressure?

Under time pressure we have to make decisions more quickly than we would like. We would like more time to plan, study, collect data, confer with others, etc. Level of stress is a function of:

- The amount of time available
- The level of risk
- Situation ambiguity
- Solution ambiguity

Since the decision must be made rapidly we must depend on previously existing knowledge to craft a creative solution in real time.

Under time pressure the tendency is to use naturalistic methods (Recognition Primed Decision method). RPD has the obvious advantage of being fast. The key to successful application of RPD is **situation assessment.**

3

Your Decision, Much at Stake, Much Time Pressure

In a situation of manageable time pressure apply the insights learned in chapter 7. In summary:

Use the time you have wisely. Realize that we tend to underestimate the time actually available. Don't rush any more than is necessary.

Focus on situation assessment. Understanding the situation is often the key to successful decision making under time pressure.

Communicate: Under time pressure communication tends to suffer. The best use of available time may be to consult with others.

Consider ways to decrease the level of risk. Decrease stress by taking action to mitigate the risk.

Identify plausible objectives. Under time pressure the optimal answer is probably not the right goal, seek a satisfactory solution..

Accept sequential search (vs. simultaneous search). Under time pressure it will not be possible to identify and evaluate a large number of alternatives.

3s

Stakeholders, Much at Stake, Much Time Pressure

Group decision making under stress is not an attractive situation in general. Apply all the insights listed under node 3 above. Under intense time pressure you will have to make the decision yourself and ask forgiveness from the other stakeholders later.

Where time permits **communicate** with others. Communication can be critical in this decision mode. But under time pressure communication tends to suffer. Others may know how to solve the problem. To the extent practical, make sure all stakeholders know the situation and know what is needed from them and expected of them.

Obviously no one wants to be in this situation. It should be avoided as much as possible via planning.

9.11 DECISION STRATEGY – ANALYTICAL DECISION MAKING

The final node is analytical decision making applied to a complex problem with much at stake and adequate time to work the issues. Any of a number of factors could be important in solving such a problem. Only one node is shown here because the impact of stakeholders should be included in the assessment of important factors. In particular, stakeholders impact what the relevant objectives are.

Agreeing on decision strategy with stakeholders: When multiple stakeholders are involved focus on the objectives and decision strategy early. Try to get all the stakeholders to agree on what the objectives are and the method by which alternatives will be evaluated against those objectives.

4

Much at Stake and Time Available

What are the important factors?

<u>Belief Generation, Situation Assessment, Sensemaking?</u> Some decisions are dominated by our evaluation of the situation. If the key to making an effective decision is situation assessment or belief generation then:

- Identify your assumptions, both explicit and implicit. What if one or more of them are wrong?
- Spend time on data acquisition if that will help clarify the situation.
- Have experts review and evaluate the data.

<u>Identifying Objectives or Alternatives?</u> In a complex situation it is likely that you will miss important objectives if you don't explicitly identify objectives and involve all stakeholders in identifying them. Failure to identify objectives may lead to failure to identify suitable alternatives.

- Explicitly identify your objectives and those of other stakeholders.
- Structure the objectives to ensure completeness.
- Use your objectives to identify effective alternatives.

<u>Selection or Implementation</u>? In many decisions it makes little difference which option you pick, but it is very important to implement the selected option well. This is especially likely if the alternative is implemented by someone other than the decision maker. If choosing the optimal alternative is the key, then focus on data collection and alternative evaluation. On the other hand, if the key to success is implementation, then focus on actionability of the alternatives. Ask :

- Which alternative has the greatest support among the people who will implement the solution? Who doesn't support it and may sabotage it?
- Who will lose a measure of control if this alternative is implemented?
- How easy or complicated will implementation be?
- Consider systems effect. Will undesirable side-effects occur?

Uncertainty? If the problem is dominated by uncertainties don't spend a lot of time optimizing the decision; tend towards the more robust, flexible options. How risk averse are you and your company?

- Evaluate objectives against the likely range of variables. Consider use of tornado diagram or Monte Carlo Simulation (Chapter 4).
- Concentrate on risk. How tolerable are the potential results if things go very badly?
- What mitigations can you implement now and/or after the fact?

Learning from previous problems? If this has been done before then learning from previous projects may be the most important feature in making decisions. If so, be careful. It is very easy to learn inappropriate lessons from past failures and to inaccurately attribute the reasons for past successes.

- Evaluate past failures and past successes in search of learnings.
- Pay more attention to successes. It is easy to learn the wrong lessons from failures.
- Consider the context. Things that work in one environment may not work in others.

9.12 DECISION STRATEGY – CLOSING THOUGHTS

Fuzzy Questions:

In developing the decision strategy over the past few sections I've tacitly assumed that most of the questions have binary answers. Of course that isn't true. These questions are fuzzy. The answers will often be in the grey area between "little" and "much" and between "simple" and "complex". Application will usually involve judgment.

The question "Whose Decision?" frequently has shades of grey. It might be mostly yours with one or more interested parties. It may not be clear how much influence any particular stakeholder expects to exert. Sometimes the other stakeholders won't care until you come up with the "wrong" answer.

Rules vs. Knowledge:

Deciding between rule-based and knowledge-based decision making can be particularly tricky. Whether an issue is simple or complex depends on your level of expertise. Experts may be able to apply rule-based decision making where novices must apply knowledge-based decision making.

Identifying Objectives:

I've added the advice "always adequately identify objectives" to the process graphic on the next page. Different features of the decision are more or less important depending on the nature of the problem, but it is always important to know what you are trying to achieve. Once again this is fuzzy advice which depends on what "adequately" means. It will mean different things depending on the nature of the problem. For time pressure problems focus on plausible, achievable objectives. For very important analytical problems seek a more complete definition of the objectives.

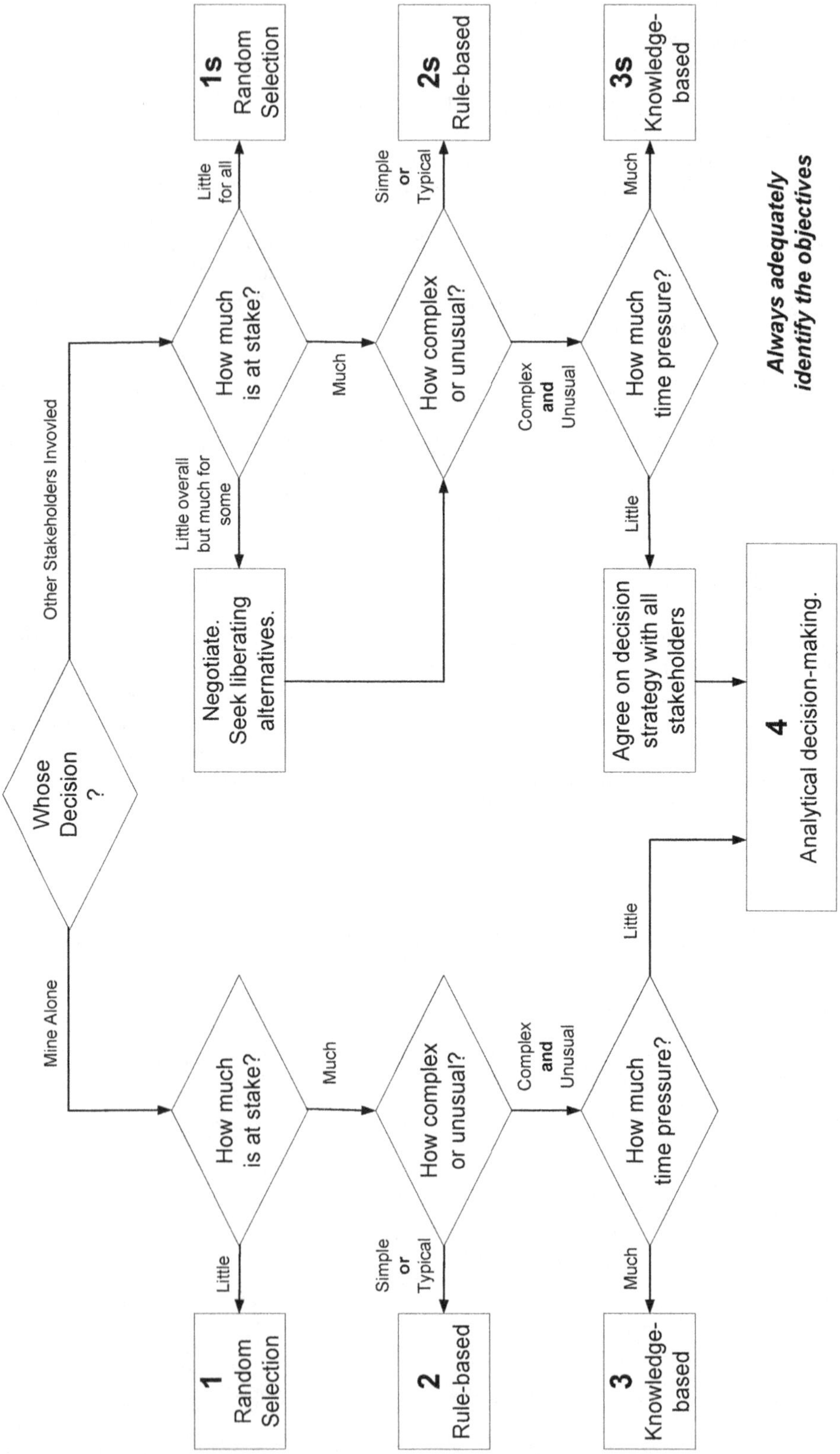
Whose Decision ?
Mine Alone
Other Stakeholders Invovled
How much is at stake?
Little
1 Random Selection
Much
How complex or unusual?
Simple or Typical
2 Rule-based
Complex and Unusual
How much time pressure?
Much
3 Knowledge-based
Little
How much is at stake?
Little for all
1s Random Selection
Little overall but much for some
Negotiate. Seek liberating alternatives.
Much
How complex or unusual?
Simple or Typical
2s Rule-based
Complex and Unusual
How much time pressure?
Much
3s Knowledge-based
Little
Agree on decision strategy with all stakeholders
4 Analytical decision-making.
Always adequately identify the objectives

9.13 CLOSING THOUGHTS

I know that there is a limited amount you can take away for a course like this.

I've written the course notes so that you can continue to study the subject, and I hope that you will. For now I'd like to reaffirm just three thoughts. If you take only three things from this course I hope that it is these:

Situation assessment is frequently key

Accurate assessment of the situation is always important. It may be especially true in time pressure situations.

Get the objectives right

To get a decision right it is crucial that you accurately and adequately identify the objectives of all stakeholders.

Identify and choose actionable alternatives

Some "good" ideas simply are not actionable. To achieve good results in any social situation you must take theory of action lessons to heart.

APPENDIX A – OBJECTIVES STRUCTURING

Objectives structuring is a powerful tool for identifying and evaluating objectives and alternatives.

A.1 FUNDAMENTAL VS. MEANS OBJECTIVES

There are two different types of objectives. Some objectives reflect what you really care about (minimizing training cost, effective training); others represent methods of achieving those objectives (using in-house experts to teach courses, using public courses).

Fundamental objectives are objectives that reflect what we really care about.

Means objectives are important only because they allow us to achieve fundamental objectives.

To determine whether an objective is a fundamental or a means objective, ask:

> *"Why do I care about this?"*
>
> If the answer is "because it allows me to achieve _____", it is a means objective.
>
> If the answer is "because it's important", it is a fundamental objective.

Structuring Objectives

Objectives structures are useful for developing, understanding and communicating objectives.

Fundamental Objectives Hierarchies

Fundamental objectives can be structured into fundamental objectives hierarchies. Not every fundamental objective has the same level of importance. For a given decision problem, one objective is likely to be the most important or overall fundamental objective. Other fundamental objectives will serve to further define or explain the overall objective.

Means Objectives Networks

We do not structure means objectives in hierarchies because means objectives are not fundamentally important. We cannot usually argue about the level of importance of means objectives. One means objective is more or less important than another means objectives only in the ways in which those two objectives impact the fundamental objectives. Hence, means objectives are organized into networks.

A.2 EXERCISE – FUNDAMENTAL vs. MEANS OBJECTIVES

We wish to improve highway safety. In the table below is a list of objectives:

Which of these are fundamental and which are means objectives?

For each objective ask: Why do I care about this?

> Because it's important ➔ Fundamental
> Because it allows me to achieve ____ ➔ Means

Objective	Fundamental or Means Objective?
Minimize fatalities	
Minimize injuries	
Maximize seat belt use	
Improve car maintenance	
Toughen car inspection programs	
Minimize property damage	
Establish curfews for teenagers	
Decrease automobile insurance rates	

A.3 EXERCISE RESULTS– FUNDAMENTAL vs. MEANS OBJECTIVES

Objective	Why is it important?	Type of Objective
Minimize fatalities	It just is	Fundamental
Minimize injuries	It just is	Fundamental
Maximize seat belt use	Seat belt use minimizes injuries and fatalities in crashes	Means
Improve car maintenance	Decreases number of crashes and minimizes injuries in crashes	Means
Toughen car inspection programs	Improves car maintenance	Means
Minimize property damage	It just is	Fundamental
Establish curfews for teenagers	Minimizes accidents caused by drunk, sleepy teenagers	Means
Decrease accident insurance rates	It just is *(but it doesn't appear to be a valid objective for this problem)*	Fundamental

Whether an objective is a fundamental objective or a means objective will sometimes depend on the context of the decision problem. For instance, if you are an employee of a state highway department and your boss instructs you to come up with a comprehensive program for increasing seat belt use in the state, then "maximize seat belt use" is the overall fundamental objective for that decision problem. In most other contexts it would be a means objective.

A.4 FUNDAMENTAL OBJECTIVES HIERARCHY

It is important to identify fundamental objectives because they capture what is important to us. But not all fundamental objectives are equally important to us. Fundamental objectives are structured into hierarchies with the overall objective at the top and lesser objectives below it.

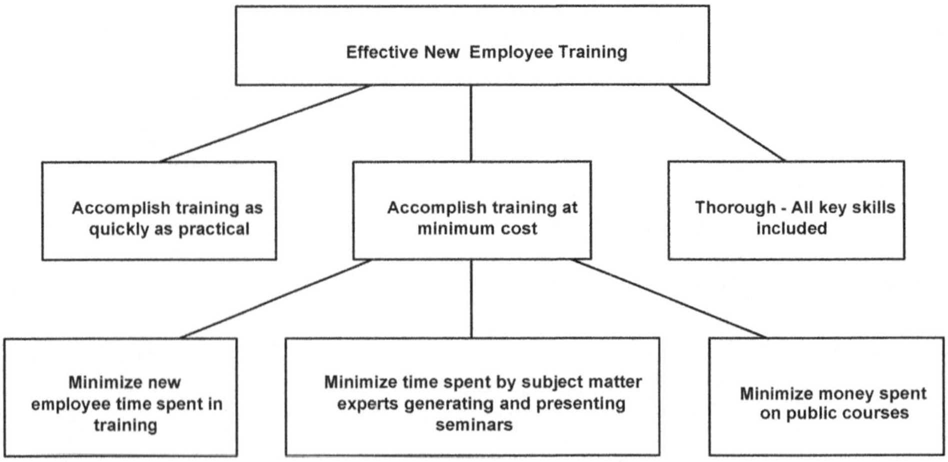

The graphic above is a fundamental objectives hierarchy for a new employee training program. The overall objective (top rectangle) is effective training of new employees. The objectives at lower levels describe what we mean by an effective training program.

Clemen (8) provides useful questions for creating a hierarchy:

To move **downward** in the hierarchy, ask:

<u>What do you mean by that?</u>

To move **upward** in the hierarchy, ask:

<u>Of what more general objective is this an aspect?</u>

Applying the first rule to the top objective:

What do you mean by "Effective new employee training"?

There are three parts to the answer, per this hierarchy:

We want to train them quickly

We want to train them at minimum cost.

And we want the training to be thorough.

It is possible now to apply the same question to these objectives and generate another level.

The question 'what do you mean by "minimize cost"?' yields lower level objectives:

"Minimum employee time spent in training"

"Minimize subject matter expert time generating courses" and

"Minimize money spent on public courses"

That question was trivial. The question "what do you mean by thorough training?" would be a very important one to answer.

Value Judgments Required - Technical Expertise not Required

Note that there is nothing technical about this hierarchy. Value judgments are required for developing the Fundamental Objectives Hierarchy, but usually little technical expertise is required. Stakeholders without technical expertise can be included in the decision making process at this stage.

A.4 FUNDAMENTAL OBJECTIVES – TESTING FOR ADEQUACY

Knowing when to stop

For a complex problem the exercise of creating an objectives hierarchy could potentially go on for quite some time as we continually refine what we mean by each objective. How do we know when to stop? It is important that we go:

High enough:	identify the right overall objective
Wide enough:	adequately identify objectives at each level
Low enough:	bottom level needs no further definition

Below is a list of desirable properties for fundamental objectives sets: (for a more complete list see Keeney (2))

High Enough

Correct Context

It is often possible to keep building the fundamentals objective hierarchy upwards to more general objectives. If we do we will, at some point, extend beyond the point where we have impact. The new employee training hierarchy above could be considered a part of a larger hierarchy including training objectives for all employees, but that larger objective may be outside of your scope.

Wide Enough

Complete and Concise

Assuming that you have the correct context (correct overall objective), next ask if the list is complete. You have to ask that question at every level. The objectives at each level should adequately describe and define all of the objectives on the level immediately above. Completeness can be carried to the extreme of being unworkably large. Compromise as necessary to maintain a concise, usable and understandable list.

Low Enough

Measurable

This is the best clue that the lowest level fundamental objectives are detailed enough. Each lowest level fundamental objective should be measurable. It is usually not necessary, nor useful to develop attributes (measures of success) for objectives above the lowest level.

We will consider attributes later in the course. Attributes are scales of measure. Some objectives have natural attributes (time, money, etc.) and are easily measured. Others do not have natural attributes and can be difficult to measure. We will discuss creative ways in which to measure objectives that have no natural attributes.

A.5 MEANS OBJECTIVES NETWORK

The fundamental objectives are the things we really care about. Means objectives are ways of obtaining the things we really care about. In developing a means objectives network we are moving a step in the direction of identifying alternatives. Indeed, each means objective can be considered as an alternative for achieving one or more of the fundamental objectives.

Factual Judgments vs. Value Judgments

Where value judgments are important in developing the fundamental objectives hierarchy, technical expertise is required for developing the means objectives network. This suggests a natural division of labor. It is entirely reasonable to have different people working on the two.

Developing a Means Objectives Networks

To develop the means objectives network begin with the overall fundamental objective and ask yourself

How would I achieve that objective?

For each answer to that question, ask the question again.

Minimizing Car Payments

The graphic below is a means objective network for minimizing the monthly payment for a new car. Minimizing the monthly car payment is the overall fundamental objective.

All of the other objectives on the graphic below are means objectives, that is, they represent means of achieving the fundamental objective of minimizing the monthly payment. None of these is important in and of itself.

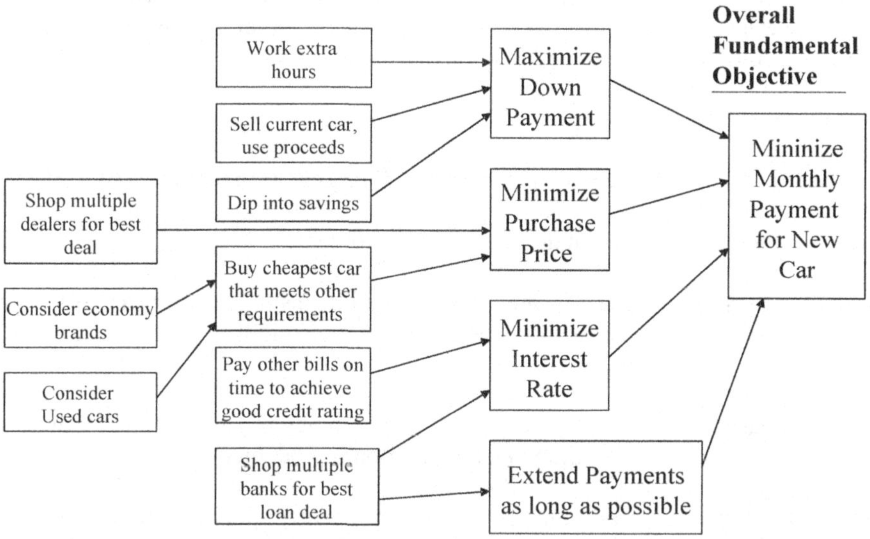

To develop the means objective network we first ask,

How would I achieve "minimize monthly payment for new car"?

The answers shown on the network are:

- Maximize the down payment.
- Minimize the purchase price.
- Minimize the interest rate.
- Maximize the payment period (number of months).

To extend the network another level, we ask the same question for each of these objectives. For example:

The question How would I achieve "maximize the down payment"? yields these ideas:

- work extra hours (to make more money),
- sell current car and use proceeds, and
- dip into savings.

A.6 ADEQUACY OF THE MEANS OBJECTIVES NETWORK

Earlier we discussed tests for adequacy of the Fundamental Objectives Hierarchy. It is more difficult to judge the adequacy of the Means Objectives Network. The Means Objectives Network describes a particular solution or a family of possible solutions. We can never know if we've missed a better solution. Still, there are some properties of a Means Objectives Network which we should value.

Fundamental Objectives Directed

The most important feature of the Means Objectives Network is that the means objectives are directed towards adequately satisfying our fundamental objectives. Hence the starting point of a Means Objectives Network must always be either the overall fundamental objective or a complete level of the Fundamental Objectives Network.

Correct Context

Are the means objectives within your sphere of influence? Can you implement them?

Is there a fundamentally different approach to the problem?

100% Ownership:	It is assumed that this will be a single owner car. If we shared use with someone else, That might cut the monthly note in half.
Buy vs. Lease:	The network presented assumes buying the car. If we leased we might get a lower monthly payment.
Car vs. Public:	Forgoing purchase and using public transportation may significantly decrease monthly expenditures, but is presumably outside the context.

Changing any of these assumptions could have a large impact on the entire network.

Complete

Complete here cannot mean what it meant for the Fundamental Objectives Hierarchy. We usually do not use all the means objectives that we identify.

"Complete" means that we've identified and selected means objectives which applied together will achieve all of our fundamental objectives.

The question to ask is "If I apply all these means objectives, will I achieve my fundamental objectives?" If not, "What's missing and how could I achieve that?"

Evaluate completeness at each node. Are the means objectives that feed into a node adequate to achieve that objective? If not, what's missing?

Actionable

We'll discuss this in much more detail in Chapter 5. Some very attractive ideas simply cannot be done because of social forces and organizational defensiveness. Ask questions like "Whose turf is this?" and "Whose support will I need?"

Effective

Often an obvious solution actually makes things worse in the long run because of unintended effects (systems effects). The US War On Drugs has probably made drug use more prevalent in the US by increasing the profit margins available to drug dealers; directly opposite of its intended effect. Teams create teamwork, but they also create turf and silos. The unintended consequences of actions is the subject of systems theory covered briefly in Chapter 5.

Internally Consistent

Can you apply all of the means objectives? Are some mutually exclusive?

A.7 COMBINING THE FUNDAMENTAL AND MEANS STRUCTURES

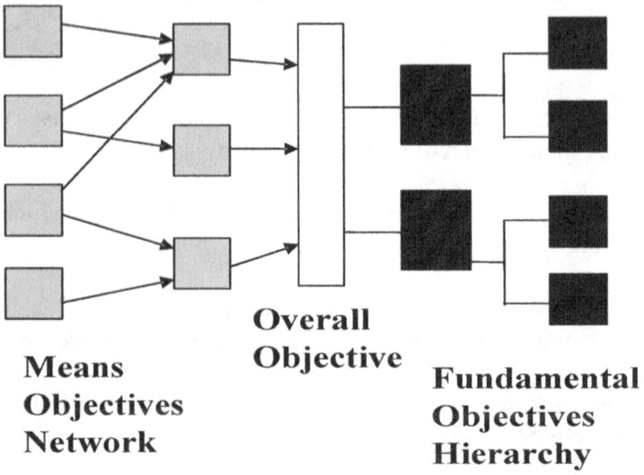

The graphic above shows the typical relationship between fundamental objectives hierarchies and means objectives networks. The classical representation has them connected at the overall objective. The overall objective is a member of both structures.

It often is clearer to connect the means objectives at a lower level of the fundamental objectives network to more clearly identify how specific means objectives impact specific fundamental objectives. There is no right way. Structure the network in a way that makes the objectives clearest to stakeholders.

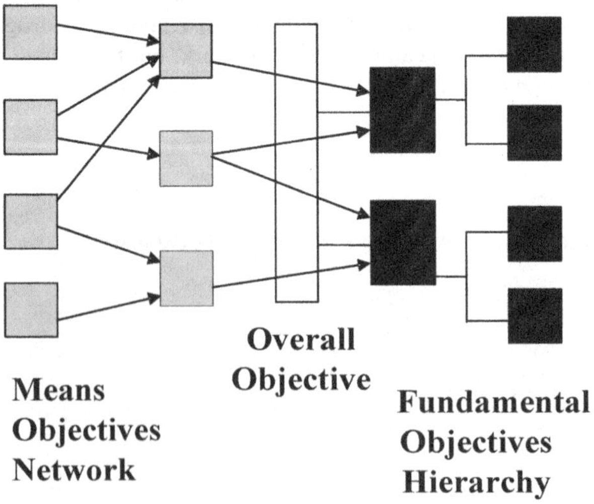

Extending the Structure

The objective structure grows in two directions out from the overall objective. We will continue expanding in both directions. This is a good time for a preview of the next steps.

Extending to the Right - Attributes

To the right we have fundamental objectives, the reasons we care. As we go out to the right from the center we are getting into more details about why we care.

Since these are things we care about, the success of the decision process will be measured by how well we achieve them. One of the criteria for a well-developed Fundamentals Objectives Hierarchy is that the lowest level objectives be measurable. The next logical step in building the structure to the right is to determine scales of measurement. We call these scales attributes.

Extending to the Left - Alternatives

To the left of the overall objective lie the means objectives. Means objectives are ways of achieving fundamental objectives. There is little difference between a means objective and an alternative. Indeed, the next step to the left is to list the alternatives.

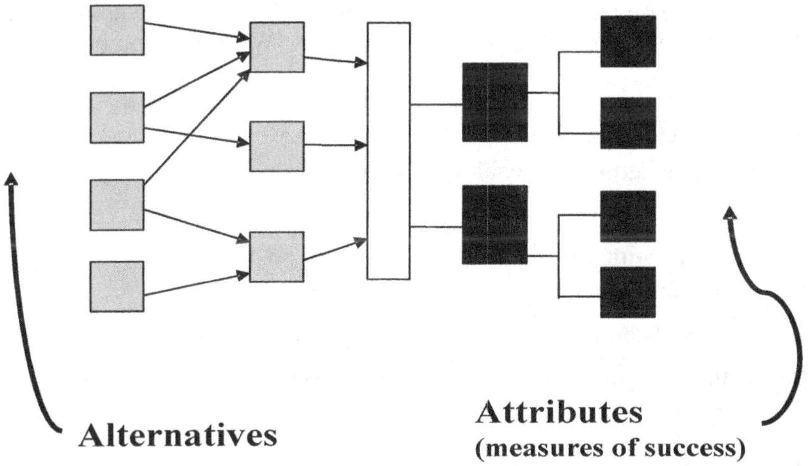

APPENDIX B - BAYES THEORY EXAMPLE SOLUTION

$$P(B/A) = \frac{P(A/B)P(B)}{P(A/B)\,P(B) + P(A/\overline{B})P(B)}$$

$P(B)$ = Probability of B occurring
$P(A)$ = Probability of A occurring
$P(A/B)$ = Probability if A occurring if B occurs
$P(B/A)$ = Probability if B occurring if A occurs
$P(A/\overline{B})$ = Probability of A occurring if B does not occur

Example Problem (from section 4.2):

Problem: A rare disease has been found in your area. 1/1000 people are believed to be ill. You have tested positive. The test is 99% accurate. What is the probability that you are ill?

B = condition of being sick
$P(B)$ = .001 (background probability of being sick)
$P(\overline{B})$ = .999
A = positive test result
$P(A/B)$ = probability of a positive test if you are sick, .99
$P(A/\overline{B})$ = probability of testing positive if you are not sick, .01
$P(B/A)$ is the probability of being sick given a positive test, which is the result we seek.

By Bayes Theory:

$P(B/A)$ = (.99) (.001) / [(.99) (.001) + (.01) (.999)] = **.09**

So, a positive test is a concern, but no reason to panic since there is still only a 9% chance that you are sick.

To get the right answer in a more intuitive way see section 4.3.

APPENDIX C – DECISION TREES

C.1 DECISION TREES

Decision trees are popular structures for evaluating alternatives. They apply the concept of expected value to make complex decisions.

DriveTek Decision
(from www.kellogg.nwu.edu/faculty/dana/htm/treeplan.htm)

You work for DriveTek, a computer storage device manufacturer. You have a contract that will pay DriveTek $200,000 for the successful development of a new tape drive. Company engineers have identified 3 methods of building a drive that will meet the customer's specifications: 1) mechanical drive, 2) magnetic drive, and 3) electronic drive. They are sure they can make the mechanical drive work (they've done it before), but it is expensive. The other drives would be cheaper to design, but there is risk that those efforts would fail. Best estimates of cost and probability of success are:

Option	Cost	Prob of Success
Mechanical	$120K	1.0
Electronic	$50K	.5
Magnetic	$80K	.7

If the option tried first doesn't work, other options must be implemented until a successful model is built. There is time to try all 3 if necessary.

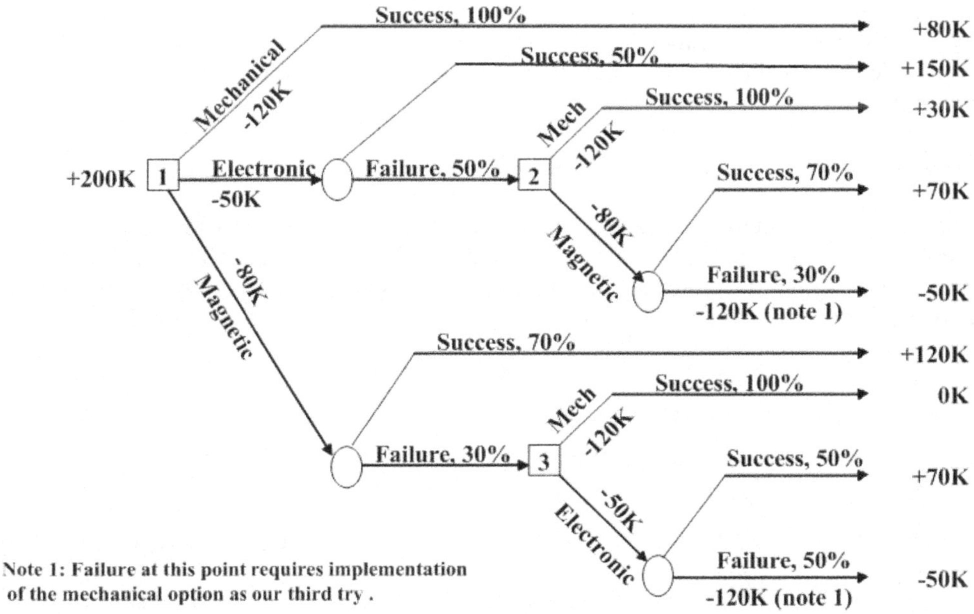

The decision tree describes the alternatives. The rectangles are decision nodes.

1	Decision node 1 represents the decision of which drive design to try first; mechanical, electronic or magnetic.

If the first choice is the mechanical option things are pretty easy. We know it will work. It will cost $120K. DriveTek will pocket an $80K profit.

The other options are more interesting. If the electronic option is selected, it may succeed (50% probability) in which case DriveTek pockets a nice profit of $150K. If the electronic option fails, there is another decision to be made at node 2.

The third leg emanating from node 1 represents the selection of the magnetic drive first. Attempting the magnetic drive will cost us 80K. If we are successful we'll make a 120K profit, if not we'll have to make another decision at node 3.

2	Decision node 2 occurs if the electronic option is tried first and fails. We have to decide which option to try next, mechanical or magnetic.

If the mechanical option is selected it will work. It will cost us another 120K and we will be left with a profit of 30K (200K – 50K for electronic try – 120K for mechanical try).

If we select the magnetic option at node 2 we will spend another 70K. There is a 70% chance we will succeed. If we do succeed we'll pocket a 70K profit. However, if the magnetic option fails we will have to implement the mechanical option as our third try. There is no decision node here because there are no other options to decide between. That will cost us another 120K and we'll wind up losing 50K on the project.

3	Decision node 3 occurs if we've tried the magnetic drive option first and failed. We now have to choose whether to try the electronic option second or go with the sure thing mechanical option second.

No surprises here. We can choose mechanical second, spend another 120K and break even. Or we can select the electronic option. Again we will be taking a risk. If the electronic drive is successful we net 70K; if not we implement the mechanical option and lose 50K.

Using the Tree

There is a great deal of information on this tree and it is presented in a logical format that allows us to see the entire decision problem at one glance. For a problem with multiple decisions, it allows us to plan a strategy. Is it clear, from just looking at the decision tree above, what decision you would make first (node 1) or what strategy you would use?

C.2 SOLVING THE DECISION TREE

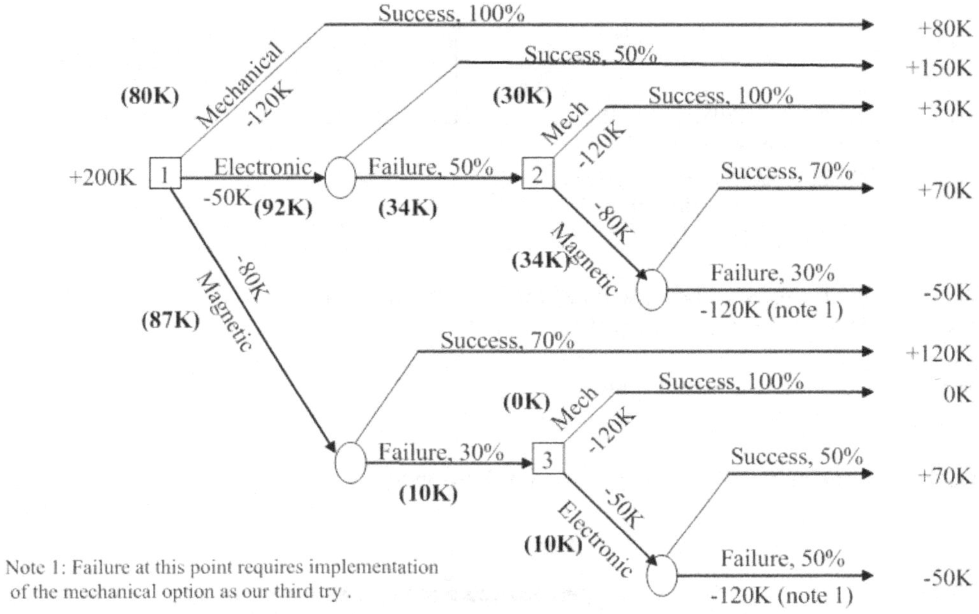

Decision trees are solved from right to left by calculating the expected value of each branch. The idea is to select, at each decision node, the branch with the highest expected value.

For example, the expected value of selecting the "magnetic" branch at decision node 2 is $34K, determined as follows:

Two things can happen on this branch:

1) The magnetic option is successful. There is a 70% chance that this will happen. If it does we make a 70K profit.
2) The magnetic options fails (prob = 30%) in which case we will implement the mechanical option and end up with a 50K loss.

The Expected Value of selecting the magnetic option at node 2 is therefore:

(Prob of success)(Value of success) + (Prob of failure)(Value of Failure) = EV

(.7)(70) + (.3)(-50) = 34K

The value of the mechanical branch at node 2 is 30K (30K profit * 100% success).

The magnetic branch has a higher expected value (34K) than the mechanical branch (30K) at node 2. We take the higher of the two and carry that value left to solve the next node. We can say that decision node 2 has an expected value of 34K.

Solving the entire tree by this method yields the following expected values for the initial decision at node 1:

Mechanical:	80K
Electronic:	92K
Magnetic:	87K

This suggests that your first choice should be the electronic drive. Further, it suggests a strategy to follow all the way to completion. That strategy is:

Actions	Success Yields	Overall Probability
1) Try the electronic option first	+$150K	.5
2) If the electronic option fails try the magnetic option next	+$70K	.35
3) If the magnetic option fails use the mechanical option	-$50K	.15

This strategy is preferred because it has the highest expected value:
EV = .5(150K) + .35(70K) + .15(-50K) = $92K

Is that the choice you would make?

C.3 CHOOSING BETWEEN RISKY OPTIONS

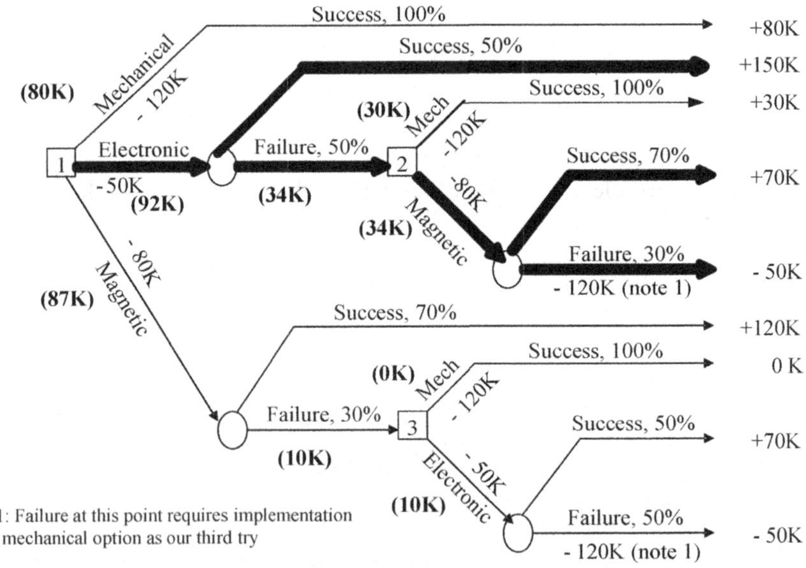

Note 1: Failure at this point requires implementation of the mechanical option as our third try

This is the branch of the decision tree we've identified as the optimal solution. We've arrived at this structure by choosing, at every decision node, the branch with the highest expected value. But we've made some risky decisions and we may lose money. Is this really what we want to do? Are these the choices we would really make?

Take a look at node 2. Here we are faced with the mechanical option (a sure 30K profit) and the magnetic option (70% probability of 70K profit and 30% probability of 50K loss)

Option		Profit	Probability	Expected Value
Mechanical		30	1.0	30
Magnetic	Success	70	.7	34
	Failure	-50	.3	

You've got to ask yourself here, "do I feel lucky"? Many (most?) people would choose the sure $30K profit here rather than risk a 50K loss.

Let's suppose that you are unwilling to accept any option that could result in a loss. Can we use the decision tree to map out a less risky strategy?

C.4 CHOOSING BETWEEN LESS RISKY OPTIONS

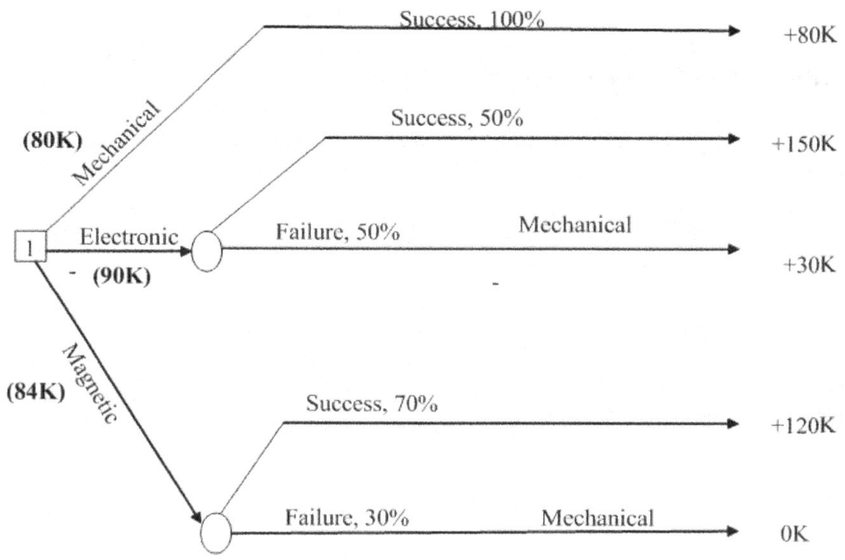

We've decided we are not willing to lose any money on this job. By eliminating decision nodes 2 and 3 (which can lead to losses) we eliminate the two riskiest options and simplify the decision tree to the structure above.

The choice is between:

Option		Profit	Probability	Expected Value
Mechanical		80	100%	80
Electronic	Success	150	50%	90
	Failure	30	50%	
Magnetic	Success	120	70%	84
	Failure	0	30%	

The electronic option is still the best initial choice by the expected value criteria.

Is that the choice you would make?

We can view the electronic option as a coin flip:

Heads you win 150K,
Tails you win 30K.

Would you play that game or take the sure 80K?

Your answer may depend on several factors such as:

Your tolerance for risk
Your company's tolerance for risk
Your company's tolerance of "failure" and its impact on your career
The size of these payoffs relative to your company's resources
Recent success or failure of similar efforts
Your confidence in the cost and probability estimates used

C.5 EXERCISE - DECISION TREE CRITIQUE

1) List the features of a decision tree that make it useful for structuring and making decisions.

2) Expected value is the decision criteria used for solving a decision tree. Discuss this approach – strengths, weaknesses.

3) In what ways does the decision tree simplify the problem? Is it meaningful to solve the problem in this simplified form?

4) Could you apply the decision tree format for any other decision criteria (other than expected value in dollars)?

C.6 EXERCISE RESULTS - DECISION TREE CRITIQUE

1) List the features of a decision tree that make it useful for structuring and making decisions.

- Complete
- Concise
- View of the entire problem space
- Useful for planning strategy as opposed to individual decisions
- Single decision criteria for comparing all options

2) Expected value is the decision criteria used for solving a decision tree. Discuss this approach – strengths, weaknesses.

- Single decision criteria applied to all options simplifies decision making.
- Does not take decision maker's risk tolerance into account.
- Requires estimates of cost and probability of success. How accurate are these estimates? Are the estimates for all three options of the same accuracy?

3) In what ways does the decision tree simplify the problem? Is it meaningful to solve the problem in this simplified form?

This tree is limited to evaluating options based on the expected value of each. Many other factors might be important including:

- Future prospects: Suppose magnetic drives are the wave of the future and this is an opportunity to gain experience on such drives. The potential loss might be well worth the risk in that case.

- Availability of resources: Presumably the people who would design the mechanical, electronic and magnetic drives work in different groups. Do all three groups have the resources required? Are the three groups equally committed to this project? Will you get the support you need from any and all groups? Suppose the magnetic group is eager to tackle this task and is actually lobbying for the job, whereas the electronics group views it as a distraction from more profitable opportunities?

The options mapped out on the Drivetek decision tree probably dramatically oversimplify the real life problem. Consider the electronic option. It is claimed that it will cost $50K with a 50% chance of success. Think about that claim for a moment; this cannot be the whole story. How realistic is it to think that we can spend $50K on an effort to develop an electronic drive and that at the end of that effort, and not before, we will be able to make an unambiguous assessment of success or failure. If the early effort is wholly unsuccessful we may be able to determine very early before spending much money at all that we will likely fail. On the other hand if we've spent the 50K and have a prototype that is very close to meeting all the specs the best option at that point might be to spend another $20K to improve the existing design rather than starting over on a magnetic option.

As the problem is stated in the decision tree the level of success apparently doesn't matter. Does success mean the same thing for all three? Presumably there are multiple criteria to be met (storage volume, speed, manufacturing cost, reliability, etc.). Different options will have different attributes. It could be that a successful electronic drive will be much more attractive to the client than a successful magnetic drive because it would be much more durable, for instance.

The problem is stated in the particular terms of cost and probability of success because those are the criteria that can be effectively managed via a decision tree. In this case it can be argued that the decision tree is used as much to define the problem as to solve it. And because of the limitations of the decision tree approach it appears that the problem, as defined, is not a meaningful problem to solve in this case.

4) Could you apply any other decision criteria in a decision tree format?

The decision tree format could be used with any decision criteria that can be specified in numeric attributes. It can also be a useful format for simply listing the available options without any attempt to solve the tree.

APPENDIX D – UTILITY THEORY

D.1 LOTTERY

A lottery is a gamble in which the prize obtained is subject to an uncertainty of known probability. The lottery depicted in the graphic below is a well-known example. As a contestant on a game show you get to choose one of three doors. You have been told that the prizes are a car, a mountain bike and a goat.

Prizes:	Car:	\$20,000
	Mountain Bike:	\$ 1,200
	Goat:	\$ 0

Expected Value

The lottery has an expected value. As we've seen before, the expected value is the sum of the value of each prize times the odds of winning that prize:

$$EV = .333 * (\text{value of car}) + .333\ (\text{ value of bike}) + .333 * (\text{value of goat})$$

$$EV = .333 * (20000) + .333 * (1200) + .333 * (0)$$

$$EV = 6660 + 400 + 0 = \$7060$$

Certainty Equivalent

The lottery also has a certainty equivalent, which is simply the amount of money you would take in exchange for playing the lottery. Imagine the game show host, after informing you of the prizes, giving you one other option: \$5000 in cold hard cash. If you accept the money, your certainty equivalent for that lottery is less than or equal to \$5000. (It is less than \$5000 if you would have taken a lesser amount). Your certainty equivalent is the lowest amount of money you will accept instead of playing the lottery.

A certainty equivalent could be more than, less than or equal to the expected value. Why do we care? The comparison of a person's certainty equivalent to the expected value tells us a something about his/her risk tolerance. A person who will take \$1000 rather than play the lottery is much less risk tolerant than a person who won't take \$10,000.

If your certainty equivalent for this lottery is \$7060 (equal to the expected value), then we say that you are risk neutral.

If your certainty equivalent for this lottery is less than \$7060 (less than the expected value), then we say that you are risk averse.

If your certainty equivalent for this lottery is greater than \$7060 (greater than the expected value), then we say that you are risk seeking.

The definitions of risk tolerance above assume that the player understands probability on some level and hence understands that the expected value is \$7060. It makes little sense to talk about risk tolerance in this way when the subject doesn't understand probability; and many people don't.

D.2 EXAMPLE RULES FROM PREFERENCE THEORY (UTILITY THEORY)

The rules of preference theory were developed by von Neuman and Morgenstern (25). They proposed eight rules to define rational behavior. We'll explore only 4 of them here to get the flavor.

I am using the term preference theory to refer to a field of study commonly referred to as utility theory. In this, I am following the guidance of Newendorp (23), and agreeing with him that the word "utility" was an unfortunate choice.

Ordering of Alternatives:

The first rule is that there is an ordering of preference for alternatives. Given alternatives A and B; you will either:

> Prefer A to B, or
> Prefer B to A, or
> Value them equally.

This seems obvious, and we would not consider a person rational who did not exhibit this property. Of course, you may change your mind tomorrow, but at any given time the above should hold.

Transitivity

This one is much debated, but except in special cases (paradoxes), we expect that:

> If you prefer A to B, and
> You prefer B to C,
> Then you will prefer A to C.

Substitutability

> If you prefer A to B, and
> you are indifferent between B and C,
>
> then we can substitute C for B and
> you will prefer A to C.

Continuity

This is more interesting and more important. Say you prefer A to B and B to C ($A > B > C$). Now consider the choice between B for certain vs. a lottery in which the prize is either A or C.

We represent the lottery mathematically via

L (A, p, C)

which means a lottery in which the prize is A with a probability p, or C with a probability of 1-p.

Consider your preferences as the value of p varies. If $p \cong 100\%$ you are almost guaranteed of winning A and have little chance of getting C. Since you prefer A to B, you will take the lottery in this case.

Conversely, suppose $p \cong 0\%$. You are almost certain not to win A and of getting C. Since you prefer B to C, you will take the certainty of B rather than play the lottery.

So, there must be a value of $p = p_0$, at which you would have trouble deciding between B for sure on one hand, and the lottery on the other hand.

For $p > p_0$ we prefer the lottery to B
For $p = p_0$, we are indifferent between the lottery and B
For $p < p_0$, we prefer B to the lottery.

These four rules along with four others constitute a mathematical definition of rationality. Much of the study of decision making since then has been the study of situations in which people violate these rules and hence are said to act irrationally.

SOURCES / RECOMMENDED READING

Books:

1) **Chris Argyris**
Flawed Advice and the Management Trap
2000, Oxford University Press

Dr. Argyris pioneered the field of action science. Action science is relatively unknown, but it contains the most important concepts in this course. Concepts from action science form the core of chapters 5 and 6. I suspect action science has never caught on with the public because much of the literature on it is not well written. Further, the ideas are rather difficult to apply, especially when compared to the business advice-du-jour which usually promises an easy solution to whatever ails you. This is Argyris' most readable and understandable book in my opinion, though friends who have read it have had trouble with the concepts.

2) **Ralph Keeney**
Value-Focused Thinking
1992, Harvard University Press

Dr. Keeney's insights and research into the importance of values in guiding the decision making process provide the basis for chapter 3 of this course. Most literature on decision making focuses on making a decision given a set of objectives and a set of alternatives. Very few entertain the concepts of setting objectives and identifying alternatives. Keeney provides a structured approach for using your values to set meaningful objectives and identify better alternatives.

3) **Gary Klein**
Sources of Power, How People Make Decisions
1998, MIT Press

Until the 1990's essentially all research on decision making was done in college laboratories. The subjects of the research were often college students with no special expertise in the subject they were making decision on (such expertise might bias the results). The subjects were usually not subjected to any meaningful stress. Dr. Klein pioneered research into how people with some expertise make decisions in stressful situations. The field is called Naturalistic Decision Making. Research in this very young field has already shown that much of the historic wisdom of analytical decision making is wrong, or at least is inapplicable in the real world. *Sources of Power* is the first book from the field accessible to the general public.

4) **David Bell, Howard Raiffa and Amos Tversky (Editors)**
Decision making
1988, Cambridge University Press

This book is a collection of papers from a veritable who's who in the field of analytical decision making. This book provided my first introductions to areas as diverse as value-focused thinking (Keeney), action science (Argyris) and impacts of disappointment and regret (Bell). It's a scholarly book. About half of the articles are readable by the general public.

5) Karl Weick
Sensemaking in Organizations
1995, SAGE Publications

Weick turns the whole subject of decision making on its head by pointing out that we often (usually?) make decisions before we have sufficient data, and then collect data to "make sense" of the decision. Many of a society's most closely held beliefs follow this pattern. If you are a religious believer, it is likely you believed before you had arguments and that you learned through study the arguments that now allow you to defend those beliefs.

6) Charles Perrow
Normal Accidents
1984, Basic Books

This book is a classic. Perrow takes a systems view of typical accidents in several different industries and shows that many accidents are the direct and unavoidable (normal) result of complicated and closely coupled technological systems. It is especially interesting to read this book after learning the concepts of action science (learning inhibitions) and naturalistic decision making.

7) Richard Farson
Management of the Absurd, Paradoxes in Leadership
1997, Touchstone

This is an incredible book. The list of paradoxes in business identified by Farson (33 in all) is simply mind-bending. They include: effective managers are not in control (action science supports this view), every great strength is also a great weakness (consider the World Trade Center), technology creates the opposite of its intended purpose. Attempting to manage people without an understanding of the concepts in this book seems futile.

8) Roger Clemens
Making Hard Decisions – An Introduction to Decision Analysis, Second Edition
1996, Duxbury Press

This is an excellent introductory level textbook on decision analysis. It is very well written and easy to understand. It includes particularly good discussions on probability and statistics, structuring objectives, influence diagrams. A wealth of examples bring the concepts to life.

9) Chris Argyris
On Organizational Learning, Second Edition
1999, Blackwell Publishers, Inc.

Try this book if you've read *Flawed Advice and the Management Trap* (1) and crave a more in depth treatment of action science. The book investigates the reasons organizations have trouble learning. Many excellent examples illustrate the concepts. This should not be the first book you read on the subject. At a minimum, understand the concept of theories of action before you pick up this book.

10) Rhona Flin, Deuardo Salas, Michael Strub and Lynne Martin (Editors)
Decision Making Under Stress – Emerging Themes and Applications
1997, Ashgate

A collection of papers from the field of Naturalistic Decision Making, the study of how people with some expertise make decisions in real-world stressful situations. A good introduction to the field; especially a good second read after *Sources of Power* (3). Contains some excellent insights on training employees for stressful jobs.

11) Antonio Damasio
Decartes' Error
1994, Putnam Publishing

Is there a fundamental error in the Descartes'' famous line "I think, therefore I am" ?

Damasio's book illustrates the importance of emotions in decision making (I feel therefore I am?). The book is a focused analysis of the case of Phineas Gage, a railroad worker who was seriously injured in an on-the-job accident. A large rod was fired through his skull when a dynamite charge exploded prematurely. Physically he made a complete recover and returned to work. Gage suffered brain damage that had no obvious effect except that it rendered him largely unable to feel emotions (anger, joy, fear, etc.). His inability to feel emotions also rendered him largely unable to make decisions in social settings. Emotions are the root of caring. We make decisions because we care.

12) Richard Lazarus and Bernice Lazarus
Passion and Reason, Making Sense of Our Emotions
1994, Oxford University Press

Emotions play a large role in our decision making, including business decisions. This is a short, easy read, yet provides a broad coverage of common emotions and a very interesting discussion of the nature of emotions and on methods of coping with them.

13) Robert Cialdini
Influence, Science and Practice
2001, Allyn and Bacon

The Hare Krishna's became rich panhandling in airports after discovering that people gave much more generously if they were first given a small gift (usually a worthless paper flower). Doomsday cults attempt to convert others to their faith more vigorously after their predictions don't come true. People value things that are scarce, even if they didn't particularly want one when they were abundant. Why? Cialdini explains all this and more in this essay on the ways people influence other people.

14) Gerd Gigerenzer, Peter Todd and the ABC Research Group
Simple Heuristics That Make us Smart
1999, Oxford University Press

Simple decision methods which use minimal amounts of data have the advantage of being fast, but at what cost in accuracy. Gigerenzer, et al have studied the behavior of simple heuristics and have come to the counter-intuitive conclusion that simple decision heuristics often perform better that very elaborate, complicated decision tools. How can this be? If the environment is simple, a simple algorithm will work effectively. If the environment is complicated, but most of the variability is contained in one or just a few variables, then a simple model which adequately captures the impact of those variables will be effective. Even when the environment suggests the need for a complicated model it is often difficult to build an adequate model and/or to collect data to effectively implement the model. The message: fit the decision strategy to the environment.

15) Diane Vaughan
Challenger Launch Decision
1996, University of Chicago Press

Investigations into the Challenger accident accused NASA of violating safety rules in order to increase production (maximize number of launches). Vaughan's careful study of the history of the accident demolishes that view. Data evaluated after the accident shows clear warning signs. But viewing this data after the accident is an instance of sensemaking (knowing the "truth" and fitting the available data to it). Vaughan painstaking shows how NASA and contractor experts used the same data prior to the crash to validate their belief that the shuttle was safe to fly (also an instance of sensemaking). Read *Sensemaking in Organizations* (5) before you read this book.

16) John Allen Paulos
A Mathematician Reads the Newspaper
1995, Basic Books

Paulos is better known for his earlier book *Innumeracy*. A Mathematician Reads the Newspaper is a similar, and I think much better book. Paulos takes simple articles out of the newspaper and shows how mathematical illiteracy (especially ignorance of basic principles of probability and statistics) on the parts of both the authors and the readers, leads to humorous and alarming errors and miscommunications.

17) Philip Howard
The Death of Common Sense, How the Law is Suffocating America
1994, Random House

A very interesting study of how overly prescriptive laws, now common in the US, are often counterproductive. Bureaucrats and lawmakers try to eliminate judgment by trying to anticipate every possible situation and to craft laws and rules that prescribe the appropriate decision in each situation. Our laws are now so complicated that no one can understand them and are continuously made more complicated as loopholes and problems are found and "fixed". But many "solutions" create other unanticipated problems. In the context of decision making, the lesson is that it is impossible to anticipate all possible situations. It is far better to focus on objectives and be prepared to make judgments in unusual cases.

18) Richard Neustadt and Ernest May
Thinking in Time, The Uses of History for Decision Makers
1986, The Free Press

One of the reasons frequently given for studying history is that people ignorant of history are doomed to repeat it. This is no doubt true, but those who <u>do</u> know their history also appear to be doomed to repeat it (witness the steady stream of errors made by State Department experts). Learning history is not sufficient. One also needs a method of applying those learnings to modern day problems. While we often encounter problems similar to problems encountered by others in the past, there are always some differences and it is seldom clear what the appropriate lessons from the past should be. Things that worked for the Romans against the barbarian hoards might or might not work for the US against drug smugglers. Things that didn't work for the ancient Jews against the Assyrians might work today for the Chechens. Neustadt and May present a methodology for applying analogies from history to today's problems.

19) Julian Jaynes
The Origin of Consciousness in the Breakdown of the Bicameral Mind
1990, Houghton Mifflin Books

The most memorable and thought provoking book I've ever read. What is consciousness? How did it develop? How long have men been conscious? According to Jaynes, consciousness is the metaphor for "I" and appeared only in the last few thousand years as languages introduced the concept "I". I didn't believe it either at first, but the arguments are persuasive and you'll believe it by the end of the book. What does this have to do with decision making? The cognitive process by which we go from "undecided" to "decided" is subconscious. This can have significant impact on what you believe and what you can defend. We frequently reach decisions (subconscious process) which we cannot logically defend (conscious process).

20) Peter Senge
The Fifth Discipline
1990, Currency Doubleday

If you've read Flawed Advice and the Management Trap (1), you should believe that most business advice is bad advice. Here's one of the exceptions. The Fifth Discipline is an excellent introduction to systems theory. Many times our actions don't achieve our desired results because we fail to take system interactions (side effects) into account.

21) Pascal Boyer
Religion Explained
2001, Basic Books

Regardless of your religious inclinations, this book should be fascinating. Boyer points out that certain concepts are common to most of the world's religions. He argues that these concepts are natural consequences of the way our brains evolved to tackle the everyday problems of living in a social society and a predatory world. Much of the book is a description of our inference systems and the ways Boyer claims these inference systems govern religious thought.

22) Peter Bernstein
Against the Gods, The Remarkable Story of Risk
1998, John Wiley and Sons, Inc.

The mathematical field of probability was developed very late historically as compared to some much more complicated mathematical fields that were developed much earlier. The reason for this late development was the inability of earlier men, guided by religious beliefs, to contemplate random behavior. For example, it was believed that when you threw a dice, God decided that it should come up "4". This is the story of the development of the field, as an aid to gamblers, against the backdrop of a culture not yet ready to accept it.

23) Paul Newendorp
Decision Analysis for Petroleum Exploration
1996, Planning Press

An excellent application of basic analytical decision methods within the oil and gas business. His discussion of utility theory, which he calls preference theory, is the best I've seen.

24) Pat Heim with Susan Golant
Hardball for Women
1992, Plume Printing

I can't recommend this book for the ideas it contains. I recommend it as a case study for evaluating business advice per the methods proposed by Argyris in *Flawed Advice and the Management Trap* (1). I first learned of Dr. Heim in a series of tapes present at the office as a diversity awareness initiative. Heim's mission is to teach women to "win" at the game of business. Her theory goes like this:

1) boys play sports - from sports they learn to be tough, to be aggressive, to play to win, to lose well, etc.
2) girls play with dolls – from dolls they learn to be nice, to play fair, to find win-win situations, etc.
3) men control the workplace and apply the "rules" of sports to the world of business
4) women don't understand the "rules"

I was intrigued by the tapes and bought her book. I found the book compelling when I first read it. But when I read it again applying Argyris' methods for identifying flawed advice I came to the conclusion that much of Heim's advice is bad advice, but for the opposite reason that most business advice is flawed. Most business advice identifies problems caused by Model I behavior and recommends Model II behavior as the solution. Heim suggests the opposite, that women's problems are the result of their Model II behavior in a Model I world.

25) John von Neuman and Oskar Morgenstern
Theory of Games and Economic Behavior
1944, 47, 53, Princeton University Press

The original presentation of utility theory. This is a frequently quoted book. Not recommended reading.

26) Ron Jeffries, Ann Anderson and Chet Hendrickson,
Extreme Programming Installed
2001, Addison-Wesley

Provides an excellent description of Extreme Programming, an apparently effective programming method which yields valuable insights into the engineering design process.

27) **Alistair Cockburn**
Writing Effective Use Cases
2001, Addison-Wesley

A description of how to write use cases. Use cases provide a structured format for describing the process of using items. They provide an excellent format for developing operating procedures.

28) **James Reason**
Human Error
1990, Cambridge University Press

An evaluation of the causes of human error. Focused on process industries. Especially interesting for its classification of error types into skill-based, rule-based and knowledge-based errors. His description of a cognitive model consisting of attentional and automatic subsystems ties in nicely with several aspects of this course.

29) **Thomas Gilovich, Dale Griffin, Daniel Kahneman (editors)**
Heuristics and Biases, The psychology of Intuitive Judgment
2002, Cambridge University Press

A collection of chapters by leading researchers in the area of heuristics. Particularly interesting is chapter 13 by Buehler, Griffin and Ross on the planning fallacy.

30) **Susan Fisk and Shelley Taylor**
Social Cognition
1991, McGraw Hill

Summary of the field of social cognition. A very easy read and a great introduction to this subject.

31) **Robert Baron and Norbert Kerr**
Group Process, Group Decision, Group Action
2003, Open University Press

Why do we seek to belong to groups? Why do we sometimes rebel? Why is rebellion punished? Why do we favor our team over others?. An excellent summary of research on teams with insights into improving performance.

32) **Donald Schon**
The Reflective Practitioner
1983, Basic Books

This book is a classic. Schon investigates the nature of professional practice. He views engineering as an interactive conversation with the environment; a conversation which informs both problem setting and problem solving.

33) Billy Vaughn Koen
Discussion of the Method
2003, Oxford University Press

A remarkably interesting book. Koen develops and defines the engineering method. We are all familiar with the scientific method. I was not even aware of the concept of an engineering method until I found this book.

34) Lipshitz, Friedman, Popper
Demystifying Organizational Learning
2006, SAGE Publications

This book is good reading. The book is very well organized, well written and easy to understand. The concept of Organization Learning Mechanisms as a focus for understanding and building learning organizations seems spot on.

35) Karl Weick, Kathleen Sutcliff
Managing the Unexpected, Resilient Performance in an Age of Uncertainty Second Edition
2007, John Wiley and Sons

In the 1980's researchers at U. California at Berkley studied organizations that operated hazardous technology, yet seemed to have fewer than their fair share of accidents. In this book, Weick and Sutcliff apply insights from their study of sensemaking to make sense of these organizations. The result is a characterization of Highly Reliable Organizations (HROs).

36) Edgar Schein
Organizational Culture and Leadership, 3rd Edition
2004, Jossey-Bass

Provides an excellent description of organizational culture, describes how it develops, identifies the role of leadership in building an organizations culture. This book revealed to me the connection between theories of action and organizational culture. With this link made, the insights from action science (chapter 5) used to understand individual behavior can also be applied to organizations.

37) Thomas Kuhn
The Structure of Scientific Revolutions
1996, University of Chicago Press

This book made 'paradigm' a household word. It's a remarkably interesting look at what science is and how it develops. Until about 300 years ago any educated person could read a scientific article or book. Now you have to be a scientist to read such things. This is because scientists can take for granted that their readers know and accept certain 'facts'. Indeed, the distinguishing feature of a field of science is that all practitioners accept certain basic premises. But occasionally experience does not match the theory. Some scientists then propose new theories and try to convince other scientists. Such scientific revolutions are difficult and can take a long time to play out.

38) James Taylor and Elizabeth Van Every
The Situated Organization, Case Studies in the Pragmatics of Communications Research

2010, Taylor and Francis

This is a fascinating book. Communication is an important part of any organization. This book addresses the features of organizational communication concluding that organizations **are** the texts that they create. In engineering texts in the form of codes, standards, processes and procedures dictate much of what we do and texts/documents are our products. I suggest you read *Sensemaking in Organizations* (# 5 above) prior to reading this book.

39) Gene Sharp
From Dictatorship to Democracy, Fourth Edition
2010, Albert Einstein Institute

The Arab Spring protests began in Algeria in December 2010 and quickly spread to other countries in the area. They seemed to be spontaneous uprising, but in fact they were years in the planning. The game plan implemented by the protesters was based mainly on this book. It is available or download. Must reading if you want to depose the dictator running your company. Also useful reading if you want to depose the dictator managing your team.

40) Jay Heinricks
Thank You for Arguing, What Aristotle, Lincoln and Homer Simpson can Teach us about the Art of Persuasion
2007, Three Rivers Press

I've read this book several times and bought at least a half dozen copies for friends. It is a very entertaining introduction to rhetoric, the Greek method of arguing. If you lack confidence in your negotiating skills, this is the solution.

41) Daniel Khaneman
Thinking Fast and Slow
2011, Farrar, Straus and Giroux

This is a great book. Kahneman summarizes his life's work, but also discusses the work of many other researchers. It is basically a history of decision theory research over the past 50 years. Especially interesting to me is the discussion of analytical decision making (slow) vs. naturalistic/intuitive decision making (fast); the main theme of the book. This is an area of long-standing conflict between the heuristics school (Khaneman, Tversky, etal.) and the NDM school (Klein, Flin, etal.)

Periodicals:

P1) Amos Tversky and Daniel Kahneman
The Framing of Decisions and the Psychology of Choice
Science, 1981, Vol 211, Jan 30, 453-458

One of the early articles on prospect theory.

P2) Amos Tversky and Daniel Kahneman
Judgment under Uncertainty: Heuristics and Biases
Science, 1974, Vol 185, Sept 20, 1124-1131

Tversky and Kahneman identified many systematic biases we hold and how these biases impact our evaluation of data and decision making. It is interesting to note that the heuristics that Tversky and Kahneman viewed as biases and sources of error are now viewed as strengths by researchers in naturalistic decision making.

P3) Thomas Gilovich and Victoria Husted Medvec
The Experience of Regret: What, When and Why
Psychological Review, 1995, Vol 102, No 2, 379-395

A thorough and captivating discussion of the impact of regret. Most interesting is the evaluation of why sins of commission are regretted more in the short term and sins of omission are regretted more in the long term.

P4) Glen Whyte and James Sebenius
Effect of Multiple Anchors on Anchoring in Individual and Group Judgment
Organizational Behavior & Human Decision Processes, 1997, V 69, No 1, 75-85

Study shows the robustness of anchoring in the behaviors of both individual decision makers and groups.

P5) Ralf Dahrendorf
Out of Utopia, Towards a Reorientation of Sociological Analysis
The American Journal of Sociology, 1958, Vol 154, Sept, 115-128

An insightful evaluation of utopian literature and a criticism of sociological theories based on utopian models. Dahrendorf disagrees with the prevailing view that societies are held together by consensus on values, arguing instead that "societies are held together not by consensus, but by constraint, not by universal agreement, but by the coercion of some by others. And as conflict generates change, so constraint can be thought of as generating conflict".

P6) Seville Chapman
Catching a Baseball
American Journal of Physics, 1968, Vol 36, P868-870
Reprinted in *The Physics of Sports*, 1992, AIP Press

Provides a theory of how baseball players judge fly balls and move into position to catch them.

P7) H.C. Sayre
The HAZOP Technique as a Structure for Operator Training
SPE 27271, Presented at Society of Petroleum Engineers Second international Conference on HS&E, Jakarta, Indonesia, Jan 25-27, 1994 (available @ spe.com)

This article gave me the seed idea for the HAZOP-based operator training program described in Chapter 7.

INDEX

Topic	Section(s)	Comments
action plans	5.1,5.2	actionable plans, effective actions
	5.3, 5.4	Exercise
	5.5	developing action plans
	5.6	identifying errors in action plans
	5.7	identifying potential for human error in procedures
	5.8	planning fallacy
	5.9-5.12	human error modeling
action science	see theory of action	
actionable plans	5.1	Definition
	5.2	Properties
	5.3, 5.4	Exercise
affect heuristic	2.10	
alternatives (identification of)	3.14	Methods
alternative-focused thinking	3.1	
Anchoring	2.3	data evaluation heuristic
	2.7, 2.8	Exercise
	2.9	Description
attentional subsystem	1.3	model of cognition
	5.9	model of cognition
attributes (measures of success)	3.5	using attributes to identify objectives
	3.4	objectives structuring
	3.9	Introduction
	3.10	Types
	3.11	constructed attributes
	3.12	proxy attributes
automatic subsystem	1.3	model of cognition
	5.9	model of cognition
Availability	2.3	
	2.5	description
bad advice	1.9	warning
	5.36	reasons why most business advice is bad advice
Bayes theory	Appendix B	
Bayesian reasoning	4.4	
belief generation	2.1	introduction
	2.2	reasons for belief (evidence, social pressure, salience)
	2.3-2.11	biases and heuristics
	2.10	risk perception
	2.11	coping with emotions
	2.12	social pressure

Topic	Section(s)	Comments
bell curve	see distribution	
biases and heuristics	2.1	
	2.3	introduction
	2.4	representativeness
	2.5	availability
	2.6	reference effect
	2.7 – 2.9	anchoring
	2.10	affect
	2.13	impact on data collection and evaluation
bounded rationality	4.25	
brain metaphor	8.13	brain as a metaphor for organizations
business advice	1.9	warning
	5.36	reasons why most business advice is bad advice
certainty equivalent	Appendix D	of a lottery
Challenger (Shuttle)	Preface	
	2.19	
Christopher Columbus	2.15	sensemaking
coordination loses	8.2	in teams
cognition	1.3	model of
	8.4	Team cognition
Cognitive Task Analysis (CTA)	7.14	description, Critical Decision Method (CDM)
	7.15	team CTA
	7.16	lessons learned studies
communication limits	8.10	Data sharing for interface management
	8.10	limits
	2.18	Pragmatics, sensemaking in organizations
Concorde crash	2.5	
confidence bias	2.4	
consciousness	1.2, 1.3	
constructed attributes	3.11	
control of attention	8.10	
Critical Decision Method (CDM)	7.14	Cognitive Task Analysis (CTA)
culture	8.3	team culture
data evaluation heuristics	(see biases and heuristics)	
decision strategy	4.24	introduction
	4.25	classes of decision strategy
	4.26	Fitting strategy to environment
	4.27	typical workplace strategies
	9.6 – 9.11	suggested decision strategy
decision trees	4.16	description

Topic	Section(s)	Comments
	Appendix C	
	Appendix	exercise – decision tree critique
descriptive (decision models)	1.7	
denial	2.11	
	5.33	
dissonance reduction	2.17	
distancing	2.11	
	5.33	
distribution	4.6	normal distribution
	4.7, 4.8	normal distribution exercise
	4.9	thinking in terms of a distribution
	4.10	normal distribution calculations via excel
	4.11	sensitivity
	4.12	Monte Carlo simulation
doomsday cult	2.12	
double-loop learning	see learning	
effective plans	5.1	definition
	5.2	properties of
	5.3, 5.4	exercise
emotions	1.4, 5.31	story of Phineas Gage
	2.11, 5.33	coping with emotions
	5.32	causes
	5.34	reference effect
engineering method	8.5	
engineering model	1.1	
	8.8	Interface management
	9.1	
error modeling	see human error modeling	
Excel functions	4.10	statistics
	4.12	Monte Carlo Simulation
	4.15	Linear Programming, optimization
expected value (EV)	4.17	Definition
	Appendix D	
Extreme Programming	6.15	
fast and frugal heuristics	4.25-4.27	
frames of reference	see reference effect	
frequency gambling	1.3	see automatic subsystem
	5.9	
fundamental objectives	3.4	definition
	Appendix A	Fundamental Objectives Hierarchy
	Appendix A	hierarchy test for adequacy
Grand Teton Dam disaster	6.11-6.14	
halo effect	2.4	
hazing	2.12	

Topic	Section(s)	Comments
HAZOP-based operator training	7.17	
heuristics	See biases and heuristics See fast and frugal heuristics	
Highly Resilient Organizations (HROs)	6.14	
human error prediction	5.7	
human error modeling	5.9	overview
	5.10	skill-based errors
	5.11	rule-based errors
	5.12	knowledge-based errors
influence	2.12	social pressure
	5.29	reciprocity rule, commitment rule
	5.30	crowd uncertainty
influence diagram	3.15	description
	1.8, 3.15	Decision Process Influence Diagram
information	2.14	Creating information from data
interface problems (answers to question 2)	2.22	Chapter 2 insights
	3.17	Chapter 3 insights
	4.29	Chapter 4 insights
	5.37	Chapter 5 insights
	6.19	Chapter 6 insights
	7.20	Chapter 7 insights
	8.15	Chapter 8 insights
	9.2	overall summary
Interfaces	8.6	Types
	8.7	key factors
	8.8	identify objectives
	8.9	data sharing
	8.10	communication
	8.11	Testing
intergroup conflict	8.2	
intuition	7.12	
late changes (answers to question 1)	2.22	Chapter 2 insights
	3.17	Chapter 3 insights
	4.29	Chapter 4 insights
	5.37	Chapter 5 insights
	6.19	Chapter 6 insights
	7.20	Chapter 7 insights
	8.15	Chapter 8 insights
	9.1	overall summary
learning	6.1	Introduction, single vs. double loop learning, action learning, systems theory
	6.2	Single-loop learning
	6.3	Double-loop learning

Topic	Section(s)	Comments
	6.4 – 6.10	TAP case study
	6.11 – 6.14	Grand Teton Dam case study
	6.14	Grand Teton Dam Disaster – HRO perspective
	6.15	Extreme Programming
	6.16	organizational learning
	6.17	bolted joint example
	8.2	in teams
Linear Programming	4.13	description
	4.14,4.15	example
	4.16	LP in Excel
lottery	Appendix D	
means objectives	3.4	definition
	Appendix A	means objectives network
	Appendix A	network test for adequacy
mental simulation	7.4	overpass rescue story
	7.6	Recognition Primed Decision Model
	7.13	
	7.17	improving performance
Model I behavior	see theory of action	
Model II behavior	see theory of action	
model of cognition	1.3, 5.9	
Monte Carlo simulations	4.12	
motivation loss	8.2	in teams
natural attributes	3.9, 3.10	
Naturalistic Decision Making (NDM)	7.1- 7.13	
negotiation	2.7, 2.8	anchoring in
	8.6	art of rhetoric
normal distribution	see distribution	
normative methods	1.7	
objective function	3.5, 3.13	
objectives identification	3.5	
objectives of the course	1.6	
objectives structuring	3.4 Appendix A	
Occam's razor	4.42	
OLMs	6.16	
optimization	4.13	
	4.14	Linear Programming
	4.15	in Excel
organization of the course	1.7	
	1.8	
organizational learning	6.16	
Organizational Learning Mechanisms (OLMs)	6.16	
overpass rescue story	7.4	

Topic	Section(s)	Comments
pattern recognition (also see similarity matching)	7.6-7.8	RPD
	7.11	Chapman's homerun
planning fallacy	5.8	
polarization	8.2	of team opinion
Phineas Gage story	1.4, 5.31	
preference modeling	4.17 – 4.23	
preference theory (utility theory)	4.17	introduction
	4.17	expected value
	Appendix D	utility theory
	Appendix D	lotteries
	4.18	value curves
	4.19	prospect theory
	4.21	regret theory
	4.23	summary
probability	4.2, 4.3	exercise
	4.5	hot streaks and coincidences
	4.6	normal distribution
	4.7, 4.8	normal distribution exercise
procedures	5.7	identifying risks in procedures
	7.18	cognitive factors in procedure following
process loss	8.2	in teams
prospect theory	4.19	
	4.22	modeling prospect theory observations via regret theory
proxy attributes	3.12	
recognition heuristic	4.25,26,27	fast and frugal heuristics
reference effect	2.3, 2.6	data evaluation heuristic
	4.19	risk tolerance effect, prospect theory
reflective practice	5.26	description
	6.10	
Recognition Primed Decision Model (RPD)	7.6	description
	7.7	situation assessment
	7.8	sources of stress
	7.10	sources of power
	7.17	improving training
regret	4.20	
regret theory	4.21	
	4.22	modeling prospect theory observations via regret theory
repeating mistakes (answers to question 3)	2.22	Chapter 2 insights
	3.17	Chapter 3 insights
	4.29	Chapter 4 insights
	5.37	Chapter 5 insights

Topic	Section(s)	Comments
	6.19	Chapter 6 insights
	7.20	Chapter 7 insights
	8.15	Chapter 8 insights
	9.3	overall summary
representativeness	2.3	data evaluation heuristic
	2.4	data evaluation heuristic
rhetoric	8.6	
risk perception	2.10	
risk tolerance (see also Preference Modeling)	2.10	affect heuristic
	4.19	reference effect on risk tolerance
	4.18	value curves
	4.20, 4.21	regret theory
salience	2.2	Reason for belief
	2.12	social pressure
	5.36	in business advice
satisfycing	4.25	decision strategy
	7.3, 7.6	under stress
schedule failure (answers to question 4)	2.22	Chapter 2 insights
	3.17	Chapter 3 insights
	4.29	Chapter 4 insights
	5.37	Chapter 5 insights
	6.19	Chapter 6 insights
	7.20	Chapter 7 insights
	8.15	Chapter 8 insights
	9.4	overall summary
sensemaking	2.1	
	2.15	examples, introduction
	2.16	properties of
	2.17	dissonance reduction
	2.18	sensemaking in organizations
	2.19	sensemaking in the Shuttle Challenger launch decision
	2.20	structural secrecy
	5.8	planning fallacy
	5.27	and relationship breakdown
sensitivity (Tornado Diagram)	4.11	
similarity matching	1.3	see automatic subsystem and pattern recognition
	5.9	
shared-mental-model team	8.4, 8.5	
single loop learning	see learning	
situation assessment	7.7	focus of Naturalistic Decision Making
Snake River Dam	See Grand Teton Dam	
social pressure	2.1, 2.2, 2.12	reason for belief

Topic	Section(s)	Comments
stochastic modeling (Monte Carlo Simulation)	4.12	
stress (chapter 7)	7.2	nature of stress
	7.3	impact (preliminary)
	7.6	Recognition Primed Decision Model (RPD)
	7.8	sources of stress
	7.9	minimizing stress
	7.17	improving performance under stress
structuring objectives	3.4	
	Appendix A	
structural secrecy	2.19	in Shuttle Challenger launch decision
	2.20	group impact
systems theory	5.13	description
sweater knitters story	2.15	sensemaking
take the best heuristic	4.25, 4.26	fast and frugal heuristics
tank farm fire story	7.5	
TAP case study	6.4 – 6.10	
teams	8.1	reasons for teams
	8.2	team basics, team productivity
	8.3	team culture
	8.4	model of team competence
	8.5	shared mental model teams
	8.8	interfaces
theory of action	1.8	
	5.1	
	5.14	introduction
	5.14	espoused theory (Model II)
	5.15	theory in use (Model I)
	5.16	Model I vs. Model II behavior
	5.17 – 5.18	exercise (Right Hand – Left Hand)
	5.19 – 5.20	exercise – Émigrés Seeking Religious Freedom
	5.21 – 5.24	exercise – Group Performance
	5.25	impacts of Model I behavior
	5.26	reflective practice
	5.27	relationship breakdown
	5.28	cognitive basis for Model I behavior
	8.3	in teams
	8.10	in data sharing
tornado diagram	4.11	
TRIZ	3.14	

Topic	Section(s)	Comments
utility theory	4.17	
unbounded rationality	4.25	
use cases	3.6	
	5.6	identifying risks in action plans via use cases
value curves	4.18	description
	4.21	regret theory
value-focused thinking	3.1-3.3	
Watergate break-in	2.9	

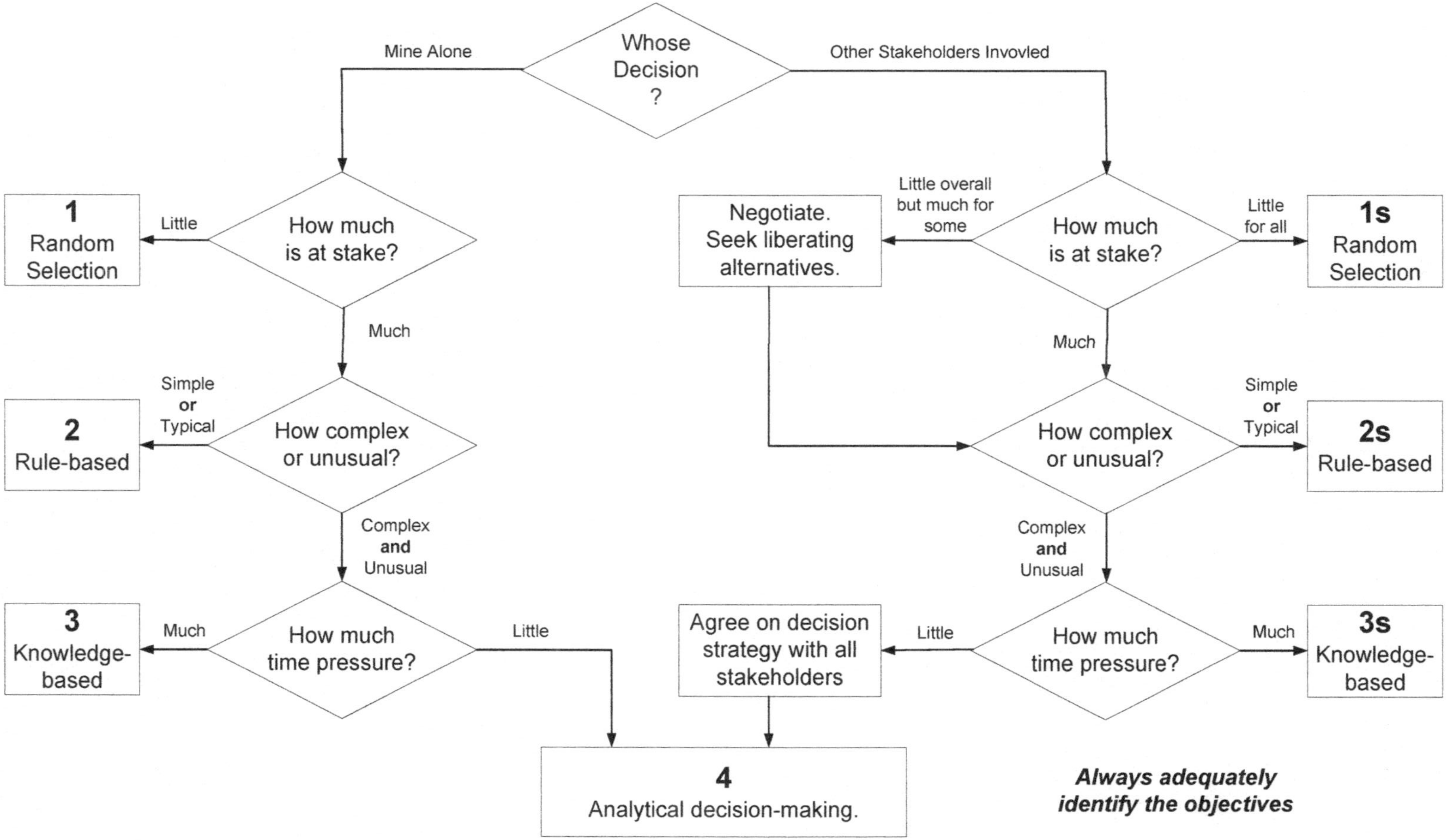
Whose Decision ?
Mine Alone
Other Stakeholders Invovled
How much is at stake?
Little
1 Random Selection
Much
How complex or unusual?
Simple or Typical
2 Rule-based
Complex and Unusual
How much time pressure?
Much
3 Knowledge-based
Little
Little overall but much for some
Negotiate. Seek liberating alternatives.
How much is at stake?
Little for all
1s Random Selection
Much
How complex or unusual?
Simple or Typical
2s Rule-based
Complex and Unusual
How much time pressure?
Much
3s Knowledge-based
Little
Agree on decision strategy with all stakeholders
4 Analytical decision-making.
Always adequately identify the objectives

Made in United States
North Haven, CT
05 November 2026

21765363R00187